155.7 Blu
Blumberg, Mark Samuel, 1961-
Basic instinct : the genesis
of behavior /
30080026564761   24.00   M

RCPL      SEP 2 7 2005

3 0080 02656 4761

Richland County Public Library
1431 Assembly Street , Columbia , SC , 29201
www.richland.lib.sc.us

# BASIC INSTINCT

*The Genesis of Behavior*

# BASIC INSTINCT

*The Genesis of Behavior*

## Mark S. Blumberg

THUNDER'S MOUTH PRESS
NEW YORK

BASIC INSTINCT
*THE GENESIS OF BEHAVIOR*

Published by
Thunder's Mouth Press
An Imprint of Avalon Publishing Group Inc.
245 West 17th St., 11th Floor
New York, NY 10011

AVALON
publishing group incorporated

Copyright © 2005 by Mark S. Blumberg

All rights reserved. No part of this publication may be reproduced or transmitted in any form or by any means, electronic or mechanical, including
photocopy, recording, or any information storage and retrieval system now
known or to be invented, without permission in writing from the publisher,
except by a reviewer who wishes to quote brief passages in connection with a
review written for inclusion in a magazine, newspaper, or broadcast.

Library of Congress Cataloging-in-Publication Data is available.

ISBN 1-56025-659-1

9 8 7 6 5 4 3 2 1

Book design by Maria E. Torres
Printed in the United States
Distributed by Publishers Group West

*FOR*
*JOSEPHINE*

# CONTENTS

What then, once again, is instinct? We see that animals feel, compare, judge, reflect, choose, and are guided in all their acts by a sense of self-love which becomes more or less enlightened through experience. It is with these faculties that they execute the intentions of nature, serve as ornaments of the universe, and carry out the intentions, unknown to us, which the creator had in forming them.

—*Charles-Georges Le Roy* (1768)

Nothing is easier to say about whatever animals do than they do it from instinct.

—*Jean-Antoine Guer* (1749)

# INTRODUCTION

INSTINCTS, IT CAN BE SAID, are everywhere.

A panicked mother rushes into oncoming traffic to save her wandering child. A salmon swims upstream to spawn. A duckling follows its mother down a garden path and into a pond. A beaver builds a dam. A young lover flies into a jealous rage at the sight of his girlfriend speaking to another boy. Thousands of flamingos gather in a salty marsh in the south of France to mate. A dog herds a flock of sheep. A green turtle swims hundreds of miles to return to the beach on which it was hatched. An interior designer displays her uncanny ability to match the finest shades of color. A newborn mouse follows an odor trail to its mother's nipple and begins to suck. A bully throws a punch in your direction and you quickly raise your arm to fend off the blow. A cowbird hatches in the nest of another species and uses its nascent wings to toss the other eggs out of the nest. Your child utters its first word.

"Many instincts are so wonderful," wrote Charles Darwin in *The Origin of Species*, "that their development will probably appear to the reader a difficulty sufficient to overthrow my whole theory." But what are instincts? We may agree initially with Darwin that "everyone understands what is meant," but it is unlikely that, on closer examination, we all would agree that each of the instances listed above qualifies as a true instinct. After all, how could any single concept meaningfully encompass herding and interior design?

A clear definition of instinct might help here and Darwin obliges us with one: "An action, which we ourselves require experience to enable us to perform, when performed by an animal, more especially by a very young one, without experience, and when performed by many individuals in the same way, without their knowing for what purpose it is performed, is usually said to be instinctive." Each piece of Darwin's definition, however, raises more questions than it answers. For example, is Darwin implying that only non-human animals have instincts? Are instincts exclusively produced by young animals? Does experience really play no role in the expression of instincts?

These are among the most basic questions that we can ask about behavior and yet experts still do not agree on the answers. Of course, we know that animals, including humans, exhibit remarkable behaviors throughout their lifetime that promote their needs to reproduce, care for and protect their young, defend territories, locate food and water, construct shelters, migrate, and communicate. Determining which, if any, of these behaviors qualify as instincts is sufficiently difficult that it really is not the case, as Darwin suggested, that "everyone understands what is meant." On the contrary, disagreement about the concept of instinct has never been more heated. In particular, the instinct concept is now being applied with renewed vigor to the study of human development by a group of scientists called nativists, who fervently believe that we are born with certain core capacities and knowledge that provide the basic structure for much of what we come to learn throughout our lives. On the other side of this debate are those who believe with equal passion that the instinct concept has outlived its usefulness and that its application to human infants by nativists only retards our understanding of human development.

The intellectual struggle with the instinct problem by numerous thinkers—from Plato to Charles Darwin to the present day—has typically revolved around a few critical questions: Is rational, intelligent thought a property of the human mind alone? If so, then how do we account for the remarkable fit between animal behavior and biological need, a fit that implies the workings of some form of intelligence? But what is the form of this intelligence—rational intelligence that springs from within the animal or divine intelligence injected from without? What is the mechanism by which offspring inherit their parents' instincts? What about experience? Is it not possible that some instincts do not simply appear suddenly, fully formed, as is commonly held, but rather develop through the accumulation of experience? And cannot this prior experience begin to have its effects in the womb or the egg, even before the animal sees the light of day?

These are not idle questions. At stake is man's privileged place in the animal kingdom and the need to posit a god as the ultimate source of intelligent design. So it is hardly surprising that virtually every major scientist, philosopher, and theologian has had something to say about instincts and their source, and much has been written about each of these individual thinker's contributions to the debate. For our purposes, suffice it to say that after hundreds of years of discussion on this topic, every conceivable intellectual niche has been filled: instinct as reflection of divine guidance and instinct as reflection of genetic guidance; instinct as a foundation for reason and instinct as the antithesis of reason; instinct as a chain of reflexes and instinct as qualitatively different from reflexes; instinct as learned behavior and instinct as unlearned behavior and so on.

The pre-Darwinian fascination with instinct as God's way of manifesting rationally intelligent behavior in otherwise irrational animals is illustrated in the works of numerous authors. For example, William Kirby's contribution in 1835 to a series of volumes known as the Bridgewater Treatises was entitled *On the Power, Wisdom, and Goodness of God as Manifested in the Creation of Animals and in Their History,*

*Habits, and Instincts*. As this title makes clear, the instincts of animals were mere subplots in the greater story of God's grand design. Other theologians took this idea even further, seeing goodly instincts as the result of angelic inspiration and sinful instincts as the result of demonic instigation.

The discourse on instinct changed dramatically when Charles Darwin not only saw the possibility of complex behavior arising without divine guidance (as imagined by many others before him), but provided a mechanism—natural selection—by which it could actually happen. Thus, it is Darwin who is typically credited with moving the instinct debate beyond the grasp of natural theologians.

Despite this rich history, here we are, having entered the twenty-first century without having attained agreement on such a fundamental issue as instinct. This state of affairs is a tell-tale sign that my field of study has not yet matured into a natural science—like physics and chemistry—in which the deep problems are well-defined and knowledge accumulates in a sure-footed manner. To be fair, particles and molecules don't exhibit the complex, highly evolved, and unnervingly variable behaviors that animals do. But that excuse can only take us so far. Sadly, the behavioral sciences remain sufficiently immature that prominent scientists disagree vehemently about such core issues as the nature of instinct and the value of nativism. To make matters worse, these scientists cannot communicate with one another because they simply do not speak the same language.

Here we will examine a variety of behaviors to show that a knee-jerk invocation of the instinct concept is not only too easy but also misleading. We will see that the term *instinct* is often merely a convenience for referring to complex, species-typical behaviors that seem to mysteriously emerge out of nowhere. But this is an illusion that is fostered by the instinct concept. Indeed, as we learn more about behavior and its development, the instinct concept loses its grip on our imagination and becomes less satisfying as an explanatory tool.

This book is not meant as a comprehensive history of the many thinkers who have engaged in squabbles over the meaning of such terms as reflex, habit, instinct, intelligence, and reason. Nor is it my aim to systematically refute the many claims that can now be found in numerous recent books on the nature-nurture debate. Instead, my goal is to provide the reader interested in the origins of behavior and cognition with a cogent perspective—a perspective that does not rely on the hackneyed explanations that have become so wildly popular among scientists and laypeople alike.

To arrive at our destination, we will discuss a wide variety of topics, each meant to illuminate a different aspect of the instinct debate and instill an appreciation for the importance of searching for the origins of behavior. Specifically, after exploring the complexities of instinct and the many problems that swirl around it (Chapter 1), we will examine our human propensity to see the biological world as the product of design and how this propensity clouds our view of instinct (Chapter 2). One popular definition of instinct involves its design and control by genes, but we will see that an accurate conception of gene function provides a more biologically plausible path to the origins of instincts (Chapters 3 and 4). Once we understand that easy appeals to genetic design are inadequate to the task of explaining instincts, we will be able to appreciate the contributions that developmental analyses of instinct have already made (Chapter 5). Armed with this background, we can then critically assess two popular trends in psychology. Specifically, we will examine the claims made by evolutionary psychologists concerning human instincts (Chapter 6) and the claims made by nativists concerning their developmental origins in humans (Chapter 7). In the final chapter, we will see how consideration of the behaviors of domesticated species, such as dogs, can provide a fresh and coherent conception of instinctive behaviors, their biological foundations, and their developmental origins.

My own journey into the realm of instinct intensified at a recent gathering of developmental psychologists (a designation reserved for those psychologists who study sensory, perceptual, motor, and cognitive development in human infants), where I attended a debate between a nativist and an anti-nativist, both highly respected researchers in the field. The packed room reminded me of question time in the British House of Commons, with members nodding vociferously when *their* person was talking, and shaking their heads and muttering when it was the opponent's turn. This experience motivated me to examine the debate in depth and to explore its parallels with a similar debate that rocked the field of animal behavior in the middle of the last century. I began reading everything I could find on the topic. But more inspiration was needed, and it soon arrived, not within the covers of a book but nestled in the cradling arms of a friend.

# 1

# A HERD MENTALITY

MILLIE WAS A PERFECT DOG. A medium-sized Hoosier hound of unknown heritage, she was stylish without being arrogant. When we walked with her about the neighborhood, she would prance, tail high, the view from the rear revealing silver-black pantaloons flowing down the backs of her legs. She rarely seemed more happy than when crawling on her belly across a patch of moist grass on a bright spring day. Millie was the kind of dog that elicited spontaneous comments from strangers who would stop their cars at intersections, even when the traffic light was green, to tell us how much she reminded them of their favorite (and usually dead) dog. Millie was also a lifetime learner: on one occasion, having successfully escaped from our yard, she took up with a graduate student and attended classes with her that day. This student was not at all pleased when we asked for Millie back.

Millie, however, was not the only dog in our household. She was, in fact, the youngest of three—the other two being older, though much smaller, Bichon Frises. After a rocky start, the relationship among the three had grown amicable over the years, and Millie had even come to allow the two puffballs to win their play fights, falling on her back while the pair of deluded Bichons hovered and growled over her. This was our family, until Millie died suddenly as a result of an enlarged heart that had gone undetected for years. We were devastated.

And then came Katy. A veterinarian friend who works with a local animal shelter came upon a six-week-old black-and-white Border collie mix who was in need of a home. Thinking of us and our recent loss, this friend delivered her to our home one day as a surprise. Despite the recency of Millie's death, this dog was simply too cute to resist, with her translucent blue eyes, shiny black coat, and velvety floppy ears. We agreed to take her in.

But Katy was a mutt of a very different kind. Almost immediately upon her arrival she began nipping at the two Bichons' heels and rumps, staring at them, chasing them, harassing them. In her puppy mind, we mused, she must have thought that her canine creator had provided her with her very own flock of play sheep. Noting the intense blue-eyed stare that Katy would adopt when planning her next assault on her unwilling playmates, we were soon referring to her as The Demon. If the Bichons could talk, they would likely have chosen a less forgiving nickname.

The autumn after Katy's arrival, while on vacation in northern Italy, we took an unplanned day trip to a region near Lake Iseo, bounded by Lake Como in the west and Lake Garda in the east—the Camonica Valley. In this valley, approximately eight thousand years ago, the end of the Ice Age brought glacial retreat that made way for new human settlement. These new inhabitants—the Camuni—thrived in this now-lush valley initially as hunters and farmers but later as traders of the products that emerged from their mining and metalworking activities.

What is most striking and unique about the Camuni is their legacy of 170,000 engravings on the faces of hundreds of rocks now contained within a modest but beautiful national park. Having left our car at the foot of the hill, we ambled up a gravel path through a forest of chestnut trees that, during our September visit, had littered the ground with their treacherous spiny seeds. The engravings—found at the top of the hill on the surfaces of rocks facing two large, imposing mountain peaks on the other side of the valley—depict many aspects of Camuni life spanning a six-thousand-year period that ends with the arrival of the Roman legion in 16 B.C.E. Whatever the motivation for these engravings, they represent a significant and diverse depiction of prehistoric life. But of all the hundreds of engravings that we viewed on our brief visit, our attention was captured by primitive images of dogs nipping at the rumps of deer, sometimes with human hunters watching nearby. Struck by the similarities between these images and Katy's incessant badgering of our two Bichons, we could not help but surmise that our modern dog was exhibiting an ancient behavior—an instinct—that was the heritage of the working dog that lurked somewhere inside of her.

I have long been fascinated with the behavior of Border collies and other herding dogs and it was this fascination that fueled my enthusiasm for raising a Border. As a young graduate student with an emerging interest in behavioral development, I watched in awe as my aunt's Polish sheepdogs ran circles around my young cousins as they attempted to ascend or descend a flight of stairs, a seemingly obvious exhibition of nascent herding abilities. Katy appeared to have similar tendencies that emerged on the staircase, although she repeatedly lunged at the Bichons rather than encircling them. Outside the house, when the Bichons were trying to do their business, Katy closely monitored their every move, often lunging at them and blocking their path as they walked about the yard. Occasionally, this insistent hounding developed into outright sadism on those occasions when Katy gripped a Bichon's neck in her jaws, a behavior that, for all I knew, is an effective technique for preventing sheep from wandering from the herd. Regardless, our recently departed Millie had never exhibited any of these behaviors—benign or nasty—so we concluded that Katy's relentless pestering of the Bichons reflected her Border heritage, regardless of how "pure" that was.

Having missed out on the first six weeks of Katy's postnatal life, I became keenly interested in the various activities in which she engaged as she adjusted to our home and gained control over her unwilling herd. One of Katy's most endearing qualities was her love of toys, and there was a period soon after her arrival when, each morning or afternoon as we congregated in the kitchen, she would roam the house and gather all of her play objects and place them in a pile on the kitchen floor. Each time that she entered the kitchen with a new toy, she would nose the others sequentially, a behavior that developmental psychologists use to infer the formation of categories in human infants. So, it seemed to us, Katy had developed the category *toy* or, perhaps more accurately, *my toy*. But was this toy-gathering behavior a prelude to herding? Do herding dogs exhibit a greater propensity to gather inanimate objects?

I raise these questions not because I will answer them—I cannot, and I do not know of anyone who can—but rather to illustrate my level of intellectual interest in this fascinating dog, regardless of how naïve that interest was. In fact, even as I engaged in my leisurely analysis of Katy's behavior, I knew that my informal observations could never lead to a deep understanding of herding behavior or, for that matter, Katy's behavior. In time, I found myself sinking deeper and deeper into facile explanations for every trivial behavior that this Border collie exhibited. The low point came during a visit to the veterinarian as I struggled to control Katy in the reception area; Katy was pulling me to-and-fro on her leash, wildly jumping up and down and scratching me when her lunges were executed in my direction. I was exasperated. Finally, I blurted to the receptionist, "She has too much Border in her." The receptionist nodded knowingly.

But the more I thought about what I had said, the more confused I became. What did I mean by *too much Border in her*? Too much what? And where?

Of course, I am fully aware of the conventional answers to these questions, namely, that Katy's ancestors were bred to exhibit certain instincts that made these dogs useful to humans as herders, flock protectors, or hunters, and the process of breeding for these instincts has resulted in the retention of various genes that program the dog's brain to exhibit these behaviors. Thus, the phrase *too much Border in her* is simply a shorthand by which I communicated the simple idea that Katy possesses breed-specific genes that have the function of compelling her to behave in breed-specific ways. Is this what I meant that day in the veterinarian's office? *Probably.* Is this what I believed to be true? *No.* You can imagine my confusion.

Wishing to replace my confusion with understanding, I began by consulting a variety of books and articles that detail the complex relationship that exists between humans and their working dogs. This relationship is ancient: Gray wolves began living in close proximity with our human ancestors as many as 400,000 years ago and new genetic research suggests that dogs diverged from wolves as many as 135,000 years ago. Domestication of dogs by humans may have occurred soon thereafter, although the evidence provided by the archaeological record does not extend beyond 15,000 years.

Some of the early domestic dogs, such as greyhounds, were bred for hunting. Others were bred for herding and protecting livestock, while still others helped in the driving of livestock from open fields to market. Much of this early domestication occurred at the hands of the Celtic tribes, and perhaps it was this Celtic influence among the Camuni peoples of northern Italy that is reflected in the engraved images of dogs nipping at the rumps of deer. Of the many Celtic tribes roaming about Europe between the first and fifth centuries B.C.E., some settled in Ireland (the word *collie* may come from the Gaelic word *colley* meaning "useful" or the word *coalie* meaning "black;" there are other possible derivations, including the Welsh word *coelius* meaning "faithful"). The further development of these early working dogs of Ireland continued elsewhere, first on the Scottish islands and later on the Scottish mainland, especially by the shepherds inhabiting the border land between Scotland and England. Thus, the *Border* collie.

Today, experts divide sheepdogs into two groups based on their behavior—guarding dogs and herding dogs—with each group represented by several breeds. Herding dogs are in turn subdivided into at least two more groups based on their tendency to round up stock and bring them toward the handler (*gathering dogs* or *headers*) or drive stock away from the handler (*driving dogs* or *heelers*). Furthermore, the herding instinct actually comprises an assortment of predatory behaviors that are thought to have been inherited from the ancestral wolf. These behaviors include orientation toward the target, showing of eye, stalking, chasing, grab-biting, kill-biting, dissecting, and consuming. Breeds, as well as individual dogs within a breed, differ substantially in their tendency to exhibit these various behaviors as well as the ease of enhancing or suppressing these behaviors through training.

At this point it had become clear to me that references to *the* herding instinct are inherently misleading, especially given the variety of herding dogs that exhibit diverse behaviors that are, in turn, differentially modifiable through training. Just at this moment of clarity, however, my confusion returned, and the source of this confusion encapsulates the many problems that arise as one seeks a satisfactory explanation for what an instinct is and where it comes from. For while there is little doubt that herding is a complex behavior exhibited by only a few specific breeds (it is amusing, for instance, to imagine a Bichon Frise herding sheep), contemporary attempts to explain these behaviors are, to put it bluntly, simplistic.

One can find numerous books that provide advice on the breeding, training, and working of Border collies and other herding dogs. The authors of these books, however, are interested in the creation of useful herders, not necessarily in the scientific exploration of the instinct concept. For example, in the book *Herding Dogs: Progressive Training,* author Vergil Holland asserts that "herding instinct refers to the desire of the dog to *do something* with the stock." Natural ability, according to Holland, is "an extension of instinct" but, curiously, a "dog with less natural ability may be more easily trained." I could understand how a dog driven less intensely by instinct might be less rigid and therefore more trainable; but I also wondered, is it not the natural ability of a Border collie to exhibit the basic components of herding that makes this dog easier to train than, let's say, a Bichon Frise? In short, Holland's formulation leaves unclear the precise relationship between natural ability and trainability.

In a related vein, in *The Ultimate Border Collie,* John Holmes defends the belief that instinct is unrelated to intelligence. This view, which has had many supporters and detractors over the years, holds that the herding instinct is no more related to intelligence than the newborn puppy's instinct to attach to its mother's nipple for milk. He writes, "There is an erroneous belief that the more intelligent a dog is, the easier it will be to train. That depends on whether it is willing to learn, in which case intelligence will be a great asset." On the other hand, "what intelligence the dog has may be completely over-ruled by an abnormally strong instinct." Unfortunately, without clear definitions of *intelligence, willingness to learn,* and *abnormally strong instinct,* there is little useful information to be extracted from these sentences. Clarity is sacrificed even further, however, when Holmes describes the occasional situation when a dog stubbornly insists on performing a behavior because "the instinct 'tells' it to do so." Given that *it* refers to the dog, we have here the notion that a dog's instinct *tells* the dog what to do. Now that is one crazy concept.

Despite this confusion, the enormous popularity of herding dogs as workers and pets has inevitably led to the construction of seemingly objective tests of herding ability. In some herding dog clubs, passing this test earns the dog the official designation "Herding Instinct Certified." Ironically, although these tests may be valuable for identifying able herders, they teach us as much about what the herding instinct is not as about what it is.

To become certified, dogs are tested on a number of dimensions, including *style* (that is, the dog's tendency to gather or drive the test herd, or exhibit some blend of the two), *approach* (that is, the dog's tendency to run wide or close to the herd), *eye* (loose-eyed dogs may have good concentration on the herd but do so with a lower intensity than strong-eyed dogs), *wearing* (a measure of the dog's tendency to exhibit the side-to-side movements that help to keep the herd together), *bark* (some dogs work silently, others don't), *grouping* (a measure of the dog's ability to keep the herd together), *power* (dogs that use excessive force lunge at the stock and sometimes grip them; Katy would score poorly on this dimension), *temperament,* and *interest.* There are also measures of the dog's responsiveness to the handler as well as a measure of the cooperativity of the test herd. These last two measures are particularly interesting because they highlight the implicit understanding of the test's designers that expression of the herding instinct is intimately connected with the dog's interactions with other animals, be they human or sheep. Oddly, it is recognized by herding experts that a dog that passes the test one day may actually prove to be a poor herder upon subsequent analysis, and that dogs that fail one day may ultimately prove to be good herders, as if the dog "suddenly has the light go on." In other words, the herding instinct is not necessarily a stable canine characteristic, thus raising the question of what it is that herding tests actually reveal about individual dogs. Given all of this uncertainty, is it not possible that *instinct* is just a shorthand to skirt complex issues and ease communication?

Ease of communication has shaped the usage of many psychological terms, and linguistic shorthands certainly have many benefits. But language can often impede understanding and, I will argue, that is exactly what has happened with the instinct concept. Thus, by now it should be clear that *the* herding instinct is more accurately described as a suite of behaviors which, depending on the handler, the dog's trainability, and even the cooperativity of the herd, can be shaped so as to create a useful herding dog. The sheer clumsiness of this description undoubtedly goes a long way toward explaining our continued comfort with such short-hands as the *herding instinct.* But no phrase, no matter how convenient, should be allowed to distract us from the complexity of the behavior that it denotes.

Even if the herding instinct is more complicated than commonly appreciated, we must still account in a satisfying way for the many components of herding behavior that, clearly, some breeds exhibit readily and some do not. For many in the dog breeding business, such a satisfying account begins with the precise documentation of the genealogical history of a dog. The linguistic similarity between *genealogy* and *genetic* is not a coincidence, and thus it is not surprising that instinct and genetics have become intertwined. Dog experts Raymond Coppinger and Richard Schneider explicitly describe this connection while also providing a metaphor to help us understand the developmental origins of instincts:

> It appears that dogs are genetically programmed to behave like dogs, but different breeds and even different dogs do not display the program in the same way. One might imagine an inherited tape-recording with all the canine motor patterns programmed on that tape: search, locomote, attach and suck are on the early (neonatal) portion of the tape, while submission and food-begging are part of the juvenile section, followed by eye, stalk, chase, grab-bite, crush-bite and dissect. These latter are perfected in the adult portion, along with scent-marking, dominance, courtship, reproductive and parental behavior. . . . Dogs do not learn these behavior patterns. They are instinctual. They appear at the appropriate time, and are directed to the appropriate environmental stimulus. . . . What a dog can learn is predetermined, governed by inherited motor patterns.

I have quoted this passage at length because it encapsulates so much of what this book aims to explore while illustrating a perspective that this book aims to scrutinize. Namely: Is behavioral development truly akin to playing a tape, and have the breed-specific behaviors of dogs actually evolved through a process that resembles the editing of the tape? Is instinctive behavior genetically programmed and, if so, does this necessarily imply predetermination? Is the division between learning and instinct as stark as suggested? And are learning and instinct the only two choices available to us for understanding the hows and whys of behavioral expression and development?

These questions are neither trivial nor easily answered. Nor are they limited to the topic of herding dogs. On the contrary, concepts such as *instinct, innate* (or *inborn*), *inherited, genetic, learned,* and *predetermined* have instigated furious debates over the nature of all animal behavior—both human and non-human—for centuries, and likely will continue to do so for many years. We can glimpse the seed of immortality of these debates in the popularity of such books as *The Language Instinct* and *The Blank Slate,* for which Steven Pinker has garnered both acclaim and scorn for his undeniably eloquent defense of nativism and evolutionary psychology, according to which much of human behavior and cognition is best understood through the lens of instinct and the evolutionary forces that shape instinct's genetic foundation. Painted in such broad strokes, there is little to dispute here. But, as with so many debates of this kind, the devil is in the details, and one detail that looms—perhaps above all others—is the simple question of what we mean by instinct. For instance, Patrick Bateson, in his review of *The Blank Slate,* chides Pinker for his vague use of instinct and provides a succinct lesson on the subject:

Apart from its colloquial uses, the term instinct has at least nine scientific meanings: present at birth (or at a particular stage of development), not learned, developed before it can be used, unchanged once developed, shared by all members of the species (or at least of the same sex and age), organized into a distinct behavioral system (such as foraging), served by a distinct neural module, adapted during evolution, and differences among individuals that are due to their possession of different genes. One use does not necessarily imply another even though people often assume, without evidence, that it does.

It is ironic that such an authority on language as Pinker would avoid precision when using the very concept that forms the core of his worldview; indeed, one will search in vain for a definition of *instinct* in *The Language Instinct.* In contrast, in Conway Lloyd Morgan's classic 1896 treatise *Habit and Instinct,* the first twenty-eight pages are devoted to "preliminary definitions and illustrations," an admirable, if somewhat boring introduction to any book. Still, Morgan appreciated how the same word can be used differently for everyday purposes (for which it is "freer and more mobile") than for scientific purposes (for which it sacrifices its mobility for a necessary "bondage" that allows for general agreement as to its meaning). But *instinct,* wrote Morgan, is a word that has its everyday meanings while having a technical meaning on which the scientists of his day were "by no means fully agreed." As Bateson makes clear, little has changed over the last century in that regard; instead, evolutionary psychologists have merely "rediscovered the malodorous aspects of instinct."

The conflation of the many meanings of instinct has become so commonplace that even authors who attempt explicit definitions of the term manage to do more harm than good. For example, in their book *The Animal Mind,* James and Carol Gould entertainingly survey the field of animal behavior and cognition through the exploration of such diverse instinct-related topics as food-begging in chicks, nest-building in wasps, navigation in bees, tool-use in monkeys, language-learning in parrots, and creativity in humans. One cannot help but appreciate these authors' ability to communicate the elegance and beauty of animal behavior in all its forms. In this sense, *The Animal Mind* is a tour de force.

A theme that is developed throughout *The Animal Mind* concerns the source of the extraordinary animal behaviors that the Goulds describe. As to these authors' view of the nature of this source there can be little doubt. In a book comprising approximately 200 pages of text, the terms *instinct, innate,* and *inborn* are used over 150 times. To their credit, the authors do attempt to define these terms, but their definitions exhibit the very kind of conflation of multiple meanings and implications that Bateson warns us about. Innate behavior, Gould and Gould write, is "behavior based on inborn neural circuits. These circuits are responsible for data-processing, decision-making, and orchestrating responses in the absence of previous experience. The 'knowledge' encoded by this genetically specified wiring is commonly called instinct." Elsewhere, they state that cognition "can be innate—passive knowledge encoded in an animal's genes and used as instructions for wiring a nervous system to generate particular inborn abilities and specializations."

Beyond these explicit attempts at definition, the authors drop their patina of precision for much of the book, which overflows with nebulous references to innate guidance, innate gestures, innate instructions, innate ploys, innate proclivities, innate criteria, innate subroutines, innate representations, innate biases, innate instincts, innate motor programs, innate avian work ethic, innate sets of conceptual rules, innate emotional displays, innate shopping lists, innate backup routines, innate neural checklists, innate micromanagement, innate food calls, innate "default" grammar, innately frightened, innately determined ends, innately specified subgoals, innately recognized consonants, innately processed vowels, inborn etiquette, inborn control, instinctive fear, instinctive channels, and instinctive repertoire. Amazingly, despite these incessant appeals to inborn capacities, despite their repeated references to preordained and prewired behaviors, and despite the seeming specificity of such phrases as *inborn neural circuits* and *genetically specified wiring,* the Goulds offer nothing by way of factual support.

Nor is there any acknowledgement in the Goulds' book that behavior develops, an omission that makes perfect sense to the committed nativist. After all, why discuss development when all of the interesting behaviors are inborn, programmed in the genes, prewired, preordained?

While I will argue that a developmental perspective is vital to any satisfying explanation of the complexities of behavior, it would be misleading to suggest that individuals who study development are more likely to eschew facile appeals to instincts and nativism. Nothing could be further from the truth. In fact, for many years now, developmental psychology has been commandeered by a highly influential group of nativists who argue that babies emerge from the womb with innate core knowledge regarding foundational principles of physics and mathematics. Clearly, then, wearing the developmental badge provides little protection against nativism.

Evolutionary change and individual development reflect two seemingly unrelated temporal aspects of our history. Because the connection between these two aspects of history is still poorly understood, it is not surprising that much of what will be discussed in this book reflects the continuing dispute concerning the contributions of evolutionary processes to our nature and individual experiences to our nurture. The recent energizing of the nature-nurture dispute in the field of developmental psychology reflects the continued hold of this dichotomous approach to development and can be attributed largely to those who are engaged in a holy struggle to reveal mental competencies in human infants that emerge without the contributions of experience. When viewed from an uncritical distance, there is nothing inherently wrong with the nativist perspective, and many talented investigators have devoted their careers to the cause. As we zoom in, however, and get a close look at the nativists' experiments and the inferences that they draw from them, some very serious problems reveal themselves (as we will see in Chapter 7).

As Bateson noted, the concepts of instinct and innate are related but not identical. Instincts need not be innate (if maternal behaviors are instincts, they are certainly not exhibited at birth), and innate behaviors need not be instincts (for example, the ability of a human newborn to recognize her mother's voice develops within the womb). But, as we will see, these concepts are confused often enough to reveal an underlying perspective that binds them together. Thus, as used by the Goulds and by some other ethologists (ethology is a subfield of biology devoted to the study of animal behavior in its natural context), instincts are units of behavior that, through their connection with genetic causes, are subject to modification by natural selection; in their hands, instincts are innate, identifiable, functional units of behavior that are clearly adaptive for the individual that "possesses" them and that are produced by specialized subcomponents of the brain (called modules). Similarly, nativists invoke innateness to identify behaviors and cognitions that reflect core capacities and knowledge that, in turn, are made possible by the evolution of genetically specified brain modules. The common themes are clear: humans and other animals possess inborn, genetically determined behaviors that are produced by highly specific neural machinery and that emerge independently of learning and other forms of experience. There is no doubt that these views of instinct and nativism have become exceedingly popular among scientists and non-scientists alike—but are they correct?

Over fifty years ago, a group of biologically oriented psychologists, inspired by the views of Theodore Schneirla and others, began an assault on the ethological perspective of instinct promoted by the Austrian-born zoologist Konrad Lorenz and his followers. The foundation of this assault was the concept of epigenesis, that is, the view that anatomical, physiological, and behavioral features arise developmentally from the continuous and inextricable interrelations between genes and the environment in which genes are embedded. Contrary to what many believe, adopting the epigenetic perspective does not merely entail the recognition that both nature and nurture are important, but rather that the dichotomy itself is meaningless, tantamount to arguing about whether hurricanes are more wind than water. At the heart of the epigenetic perspective is the notion that all experience—from the chemical environment of the first embryonic cell to the social environment in which the organism develops and lives—is essential for the journey from fertilized egg to fully realized organism. Although a few ethologists took the epigenetic criticism to heart and performed experiments showing that some iconic instincts (such as pecking in newly hatched chicks) are altered by developmental experience, and although psychologists have provided powerful demonstrations of the reliance of some instincts on prenatal or prehatching experience (as we will see in Chapter 5), many ethologists pay only lip service to these issues and continue to invoke the concept of instinct as an innate, hardwired, programmed, and immutable form of behavior.

The nature-nurture dichotomy has proven especially attractive and enduring, but it is only one of many such dichotomies that has shaped our current understanding—and misunderstanding—of behavior and cognition. Here is a brief list:

| | | |
|---|---|---|
| Animals | vs. | Humans |
| Instinct | vs. | Reason |
| Unlearned | vs. | Learned |
| Innate | vs. | Acquired |
| Genetic | vs. | Environmental |
| Biological | vs. | Psychological |

While the entries within each column may seem to bind naturally to one another, reflecting once-popular as well as still-popular perspectives, we will soon see why these dichotomies are false and how poorly the entries within each column actually match one another. Here is a cursory, non-exhaustive appraisal of the dichotomies listed above to show how the items across the two columns are not truly separable: Humans are animals; if non-human animals have instincts, then so do humans; genes and other biological factors play as significant a role in the construction of learned behaviors as they do in the construction of unlearned behaviors; some knowledge is acquired through prenatal learning and is therefore innate; characters that are innate can be environmentally induced; and the division between biological and psychological processes reflects more the institutional organization of universities than the actual organization of animals. Nonetheless, the allure of these and many other dichotomies persists, placing a relentless drag on any attempt to move beyond them toward a more accurate, sophisticated, and nuanced appreciation of instinctive behaviors and their developmental origins.

So where do instincts come from? Answering this question requires first that we resist the expectation that adult instincts, like nested Russian dolls, can be identified in miniature form in the developing animal. In other words, development is more than mere growth and maturation. Instead, a more fruitful path to instinct can be found by adopting the epigeneticist's broad definition of experience and accepting development as a nonrational process in which adult behaviors can emerge from nonobvious sources. For example, the so-called maternal instinct of female rats to retrieve a straying pup can be eliminated by rearing the mother with powdered food, thereby depriving her of the normal experience of handling objects (such as food pellets). Even more bizarre is the finding that the profound fear of snakes exhibited by Japanese macaques, considered an instinct in monkeys as well as humans, relies on the monkey having gained early developmental experience handling and eating live insects; or that newly hatched chicks that are prevented from seeing their toes are less likely to consume mealworms! Few such examples exist because they are so strange—so nonobvious—but the fact that they exist at all should give pause to those of us who expect to find simple, rational, obvious, one-to-one relationships between instinctive behaviors in adults and their developmental precursors in infants.

Inherent in the rationalist perspective of instinct are the notions of purpose, goals, and design. For humans, these are deep-seated, almost reflexive notions that are not easily inhibited, especially when we witness firsthand those exquisite animal behaviors that seem so perfectly designed to achieve the imperatives of survival and reproduction. Our propensity, however, to view the natural world and all of its wonders through the lens of design has only hindered our understanding of biological complexity, including behavior, as we will now see.

# DESIGNER THINKING

I WAS HOME FROM COLLEGE for the winter holidays. One evening I had settled into a comfortable recliner to read an old paperback on logic that I had discovered among my father's collection of books. It was cold outside and the house was noticeably still. I was alone, or so I thought. As with most books on logic, the topic of human rationality had to be broached, which this one did by noting that "we are all familiar with the definition of *man* as a rational being." At that moment, I heard a sound and looked up. I was staring into the barrels of two snub-nosed revolvers aimed directly at my head. The rational beings holding the guns were wearing ski masks. One of them ordered me to the ground, whereupon he bound my hands behind my back and placed my head face-down on a soft pillow. Any comfort that I may have felt was quickly dissipated when one of the gunmen suddenly turned up the volume of the television, releasing an explosion of sound that convinced me that I had but a few seconds left to live. Thankfully, my guests were thieves, not murderers. Nonetheless, at that moment I forever lost interest in what logicians have to say about human behavior.

Our commitment to the notion that humans are rational is as old as philosophy itself, inextricably bound to an equally ancient need to asso-ciate ourselves with a higher being and distinguish ourselves from mere animals. Humans, according to this still-popular perspective, possess souls, which are the vessels that transport us from this life to the next. Deserving transport to salvation, however, requires all of us to choose correctly between good and evil, and this choice is made through the exercise of reason, a capacity that only we and God possess. Animals, on the other hand, are not members of this elite club, lacking souls and, therefore, the potential for an afterlife; without souls, it is argued, ani-mals have no need for reason.

It has long been recognized, however, that animals cannot be so easily dismissed from the realm of rationality; indeed, they possess behavioral capacities that are reminiscent of the faculty of reason. How, then, many have wondered, can we account for these capacities without elevating animals to our level? The easy and comforting solution has long been to say that animals behave out of instinct.

Long before Darwin, philosophers, naturalists, and theologians noted and struggled to explain the complex behaviors of animals that, seemingly in the absence of learning, allow them to survive the chal-lenges of the natural environment as if designed to do so. When Galen, the second-century Greek physician, extracted an infant goat from its mother's womb and placed before it several bowls containing such liq-uids as milk, wine, and water, he tells us that the kid chose the milk. Galen did not conclude that the kid was behaving rationally, but that it behaved as *if* it were rational. This form of rationality, however, was thought by Galen—and many others before and after him—to be the gift of a beneficent creator. For how else were such thinkers to explain the natural wisdom expressed by such lowly creatures?

The concept that we call *instinct* has had a long and convoluted his-tory. Part of the difficulty in tracing the history of such a complicated idea is that language, culture, religion, philosophy, and science interact with such turbulence that one can rarely trace with confidence its trajectory over time. Knowing the etymology of a word—for example, that *instinct* is derived from the Latin word *instinguo,* meaning "to excite or urge," and that it is related to the word *stimulus,* a contraction of *stig-mulus,* an "object that was used to prod mules"—provides some sense of place but, ultimately, does not satisfy. Etymology may inform us that the originators of the word *instinct* used it to denote the urge to act, but it cannot tell us whether that urge in fact originates in the mind of God or the mind of an animal. What we can say with some confidence, however, is that instinct has been repeatedly employed throughout history as a means of erecting an unbreachable wall between rational man and unthinking brute.

Reason and instinct. For many, these two terms are complete opposites, one denoting the freedom that empowers the human mind and the other the shackles that doom animals to a life of automatism. Despite their differences, however, reason and instinct reduce to a single concept that, when fully appreciated, provides a foundation for understanding many features of biological history across many dimensions of time and space. That concept is design. As we will see, designer thinking has permeated, and continues to permeate, topics as diverse as religion, evolution, mind, and human invention. Understanding the attractions and pitfalls of this form of thinking is essential if we are to understand the nature and origins of instincts.

### *The Argument from Design*

A few years ago I attended a public lecture by a scientist who had written a book proclaiming that Darwin was wrong. The author of this book, Michael Behe, was promoting what many in the audience believed to be a new idea: that animals are the product of *intelligent* design and not, as Darwin argued, the product of *apparent* design. That Behe is a biochemist seemed to lend credence to his argument, as he shrewdly spoke over the heads of his non-scientific and predominantly religious-minded audience and wowed them with the complexities of life in its most miniature forms. Consider the bacterium, he preached, with its flagellum designed so magnificently for forward propulsion, like an outboard motor with all of its parts working interdependently to accomplish its function. Such interdependency, he continued, could no more be the product of gradual, blind evolution than a mousetrap. Each part of the mousetrap has a function only within the context of the complete device; remove the spring and the mousetrap is rendered completely—not only partly— useless. Having deftly maneuvered his audience to this point, he was ready to complete the bait and switch: If an evolutionary explanation fails for a mousetrap, then how could it possibly not fail for a flagellum? To bring home his point, Behe projected onto the screen a mechanical drawing of the flagellum and its associated apparatus that looked like it had been removed from the desk of the mechanical engineer who had designed it. Nice show. The audience bought it.

And why wouldn't they. After all, this argument—the argument from design—has proven its rhetorical effectiveness for centuries. Although little more than an appeal to analogies between human and natural con- trivances, the argument from design provided an important intellectual foundation for the religious faith of scientists and theologians alike, with Plato, Thomas Aquinas, and Isaac Newton among its many advo- cates. Three hundred years ago, Newton invoked the argument in his masterwork *Principia,* thus providing the imprimatur of the greatest sci- entist of his age. With such an ancient and esteemed pedigree, it is not surprising that it took an iconoclast like the English philosopher David Hume to write what many consider to be the definitive refutation of the argument from design, the *Dialogues Concerning Natural Religion.* Hume knew that he was addressing a touchy subject so, despite his iconoclasm, he arranged for his work to be published posthumously. As it turned out, Hume completed both the *Dialogues* and his life in the same year—1776.

Hume wrote the *Dialogues* from the perspective of three primary characters: Cleanthes, the scientific believer; Demea, the orthodox believer; and Philo, the skeptic (presumed by many to be the voice of

Hume himself). It is left to Cleanthes to enunciate the argument from design, which he does with great eloquence:

> Consider, anatomize the eye; survey its structure and contrivance, and tell me, from your own feeling, if the idea of a contriver does not immediately flow in upon you with a force like that of sensation. The most obvious conclusion, surely, is in favor of design; and it requires time, reflection, and study, to summon up those frivolous though abstruse objections which can support infidelity. Who can behold the male and female of each species, the correspondence of their parts and instincts, their passions and whole course of life before and after generation, but must be sensible that the propagation of the species is intended by nature? Millions and millions of such instances present themselves through every part of the universe, and no language can convey a more intelligible, irresistible meaning than the curious adjustment of final causes. To what degree, therefore, of blind dogmatism must one have attained to reject such natural and such convincing arguments?

Cleanthes's choosing of the eye as definitive evidence for the creative hand of God was an obvious one. Indeed, explaining the origin of the eye's seemingly extreme perfection would prove to be a challenge to Darwin's theory of natural selection and, ultimately, one of that theory's greatest successes. As Darwin acknowledged in *The Origin of Species,* "To suppose that the eye with all its inimitable contrivances for adjusting the focus to different distances, for admitting different amounts of light, and for the correction of spherical and chromatic aberration, could have been formed by natural selection, seems, I freely confess, absurd in the highest degree."

Consistent with his rhetorical style, Darwin erects a seemingly impassable barrier and then easily bounds over it by outlining the process by which "numerous gradations from a simple and imperfect eye to one complex and perfect can be shown to exist, each grade being useful to its possessor, as is certainly the case." For example, we know of animals with mere patches of light-sensitive cells on the skin surface; animals with indentations on the skin surface that contain light-sensitive cells; animals in which these indentations are enlarged to form eye cups that direct light to the light-sensitive cells; animals in which the eye cup has closed to form a pinhole so that light can be focused; animals with eye cups containing a gelatinous substance that acts as a crude lens; animals, like us, with a more refined lens, as well as the ability to adjust the amount of light entering the eye. Thus, eyes did not evolve through the insertion of ready-made parts, but rather evolved such that even the most primitive eyes found throughout nature benefit the organisms that possess them. Better to see a little than not at all.

Although Darwin judges the human eye an organ of extreme perfection, we know that it is not. For example, what we call the blind spot can be reasonably described as the result of a design flaw, produced by a developmental wiring problem that any thoughtful designer, working from scratch, would have avoided. Hume may not have had such detailed biological evidence at his disposal, but he was acutely aware of the "inaccurate workmanship . . . of the great machine of nature." Taking this argument to its logical extreme, Philo ridicules the argument from design by noting the many imperfections to be found in the world. Thus, for all we know, our world "was only the first rude essay of some infant deity who afterwards abandoned it, ashamed of his lame performance." In time, this argument from imperfection would provide perhaps the most convincing evidence for evolution and against intelligent design. As the late evolutionary biologist Steven Jay Gould once noted, "Odd arrangements and funny solutions are the proof of evolution—paths that a sensible God would never tread but that a natural process, constrained by history, follows perforce."

Hume's *Dialogues* seriously wounded the argument from design by revealing its logical flaws. Then, also in 1776, the Scottish philosopher Adam Smith published *The Wealth of Nations* and dealt the argument a further blow. Smith's contribution was the notion that economic order can emerge when each individual is free to behave without constraint, as if an invisible hand (to use Smith's metaphor) were molding the order according to a grand design. Inspired in part by Smith's ideas, Darwin provided the knockout punch to the argument from design in *The Origin of Species* by providing a mechanism—natural selection—by which order can arise without thought, mentation, intelligence, or design.

Organic evolution plays out on a grand temporal scale—thousands and millions of years—and it was the cloak provided by the vastness of time that obscured the mechanisms of evolutionary change and enhanced the illusion of intelligent design by an unseen creator, that is, an invisible hand guided by an invisible mind. The direct link between God and mind was apparent to Hume who, in the words of Philo, asks, "And if we are not contented with calling the first and supreme cause a GOD or DEITY, but desire to vary the expression, what can we call him but MIND or THOUGHT, to which he is justly supposed to bear a considerable resemblance?"

If the contrivances of nature, once imagined to be the product of a heavenly god, could be successfully moved to the realm of blind mechanism—of evolutionary trial and error playing out across the immensity of time—what about the contrivances of human beings, imagined to be the product of a creative, rational, purposeful, but more earthly mind? Is the man-made mousetrap evidence of the flagellum's intelligent design, or is a flagellum evidence of the evolution of the mousetrap? Before making the leap to humans, however, let's consider the role of trial and error in animal behavior.

### Cat and Mouse

A rational mind lurking behind human contrivances is the foundation upon which the argument from design rests. It is a given. Two centuries ago, William Paley famously invoked the argument from design in his *Natural Theology, Or, Evidences of the Existence and Attributes of the Deity, Collected from the Appearances of Nature.* In that book, he contrasted the obvious simplicity of a stone with the equally obvious complexity of a watch to make the point that all complex objects must have a designer, whether it be a watchmaker or God.

Paley's error is simple yet profound. When we don't directly observe the developmental origins of any historical process, we naturally gravitate toward explanations that make the complex seem simple. Such explanations, however, are mere illusions. To avoid Paley's error, the first step is to ask: Where did *that* come from? The next step is to never stop asking that question.

As already mentioned, Darwin's theory of natural selection broke the spell of the argument from design by providing a natural path to the origins of organismal diversity on our planet. This radically new perspective of life and time removed thought and planning from a process that had been assumed to be the product of a divine mind. But does Darwin's perspective translate to other time scales? For example, what about animal behavior, which unfolds in seconds and minutes?

Toward the end of his career, Darwin eyed a successor, George Romanes, whom he hoped would apply the elder scientist's evolutionary ideas to animal behavior. The study of animal behavior was in its infancy at the end of the nineteenth century, relying to a large degree on the informal observations of amateur naturalists living and traveling throughout the world. Romanes collected and disseminated these often fanciful anecdotes as facts. For example, one anecdote recounted by Romanes gained credence by its being reported by several independent observers in Iceland. It was reported that mice work together to load provisions of food onto cow paddies, whereupon they launch the paddies, hop aboard, and navigate across the river using their tails as rudders. How could mice develop such extraordinary nautical skills? Perhaps, Romanes thought, the mice had observed humans loading provisions onto boats and steering those boats with a rudder. All that was required from the mice, then, was the ability to imitate human behavior. Of course, mice have no such ability, and this fanciful story of marine-going murines has become a cautionary tale about the dangers of anecdotes; nonetheless, the story buttressed Romanes's conception of imitation as a primary source of novel behaviors.

Romanes's belief in the power of imitation is illustrated most famously in an anecdote concerning a cat that belonged to his coachman. Romanes had noticed the cat using its front paws to unlatch the lock on the gate at the front of the house. But how could a cat perform such a feat? Romanes arrived at a simple conclusion. He imagined that the cat reasoned, "If a hand can do it, why not a paw?"

While the anecdote of the Icelandic mice illustrates the unreliability of amateur observations, the ability of the coachman's cat to unlatch a gate has been observed countless times by countless cat owners. Romanes's explanation for both anecdotes, however, is the same: imitation. Romanes was attracted to imitation because it seemed to him like a more simple and reasonable explanation than the two possible alternative explanations, namely, instinct and learning.

The reliance of Romanes on anecdote and unbounded conjecture fueled one young experimental psychologist, Edward Thorndike, to perform an experiment that simultaneously ridiculed Romanes and helped to found the scientific study of animal learning. For his experiments, published in 1899, Thorndike placed hungry cats inside small "puzzle boxes" of his own design, each version of the box requiring a unique set of actions that, when performed, opened the box and freed the cat to exit and receive food and water. After escaping, the cat was placed back in the box and tested again.

Thorndike made two important discoveries from this simple experiment. First, he found that the time required to escape from the box decreased with each successive test, indicating that the cats were learning. But second, he found that cats learned to escape through a process that had nothing to do with reflection and imitation. Naïve cats met confinement in the box with agitated and random behavior that occasionally, by happenstance, produced actions that opened the box. Over time, the cats' random agitation diminished and their behavior became increasingly directed toward the latch; eventually, the cats' behavior was so focused and efficient that, like Romanes's coachman's cat, the final form of the behavior *appeared* purposeful and human-like. Thorndike's point, of course, was that the purposefulness of a behavior can cloak the aimlessness of its origins.

The trial-and-error learning process introduced by Thorndike has been compared to natural selection. In trial-and-error learning, however, behavioral variability within an individual is reduced by weeding out the useless behaviors so that what remains, after many trials, is the elegant solution. In other words, the unfit behaviors are not reproduced in successive learning trials. And just as with natural selection, when we view the final product without having witnessed the process that led up to it, we are easily fooled into thinking that intelligent design is at work.

But does a trial-and-error explanation of some aspects of cat behavior provide any insight into the rational and purposeful behaviors of human beings? Do the behaviors of a trapped cat teach us anything about the invention of a mousetrap—about human creativity and ingenuity?

### Failing to Succeed

Sitting in that audience a few years ago, listening to that modern purveyor of the argument from design, I was struck by Behe's confident assertion that human artifacts—clocks, computers, CD players—are created through a rational, thoughtful process. Of course this assertion is hardly new: recall William Paley's analogy between watchmaking and worldmaking. So, at the conclusion of his talk, I raised my hand and asked the speaker if he was aware of the possibility that even human artifacts evolve through a process of trial and error? The blank look on his face answered my question.

Henry Petroski is an engineer who has written extensively about the evolution of human artifacts. By examining in detail the history of such common, low-tech objects as pencils, paper clips, forks, and zippers, he provides a perspective that runs counter to the romantic image of the lonely inventor creating novel devices de novo using little more than reason and inspiration. On the contrary, according to Petroski, invention often is a trial-and-error process in which each successive development of an artifact is achieved through the removal of those features that don't work—those irritants that prevent an artifact from being as useful as it can be. Like evolution, design entails the removal of the unfit, producing in the end an artifact that appears to be the product of forethought, of intelligent design. As Thomas Edison famously expressed it, inventors fail their way to success. As Petroski expresses it, form follows failure.

Petroski demonstrates this evolutionary view of human invention through numerous examples. For instance, one may marvel at the beauty of the Brooklyn Bridge and see it as a stand-alone creation, but the reality is that, from the moment that the first man chopped down a tree and laid it across a creek, bridges have evolved through a trial-and-error process that has included numerous spectacular failures. Indeed, these failures are essential for future innovation. The same can be said for any of the great structural, architectural, or mechanical achievements, from the pyramids to the great medieval cathedrals to the space shuttle. One need not look to the great engineering achievements, however, to gain a sense of the evolutionary forces at play.

Consider a dining room table set for a formal function, replete with a variety of eating utensils: dinner forks, salad forks, dessert forks, carving knife, cheese knife, steak knives, fish knives, butter knives, dinner spoons, soup spoons, etc. For the most part, each utensil seems well-suited, perhaps even ideally suited, to the food for which it is intended. Upon observing such a fit between form and function, one might imagine that these utensils were designed for their purpose by a small group of particularly insightful food fanciers. Not so, says Petroski.

The diversity of eating utensils can be traced to a co-evolutionary relationship between the knife and the fork (the spoon has a separate, but no less interesting, history). This history is convoluted, but it follows a logical progression and is shaped by a few fundamental needs. The first need is basic: getting food from the plate to the mouth without having to touch the food with one's fingers. Because many non-Western cultures consider it acceptable to use fingers when eating (to compensate, Africans and Arabs have developed more rigorous cultural norms concerning hand-washing before and after meals), we are discussing here a tradition that arose primarily in Europe and was transported later to other parts of the world.

We begin with the knife, originally adapted from sharp pieces of flint. Over time, knife-making skills improved and, eventually, bronze and iron replaced stone. Knives were used for many things, including personal defense and, eventually among the more refined, eating. But how does one eat a piece of meat with nothing but a knife? Early on, bread was used to steady the meat for cutting, whereupon the meat was jabbed with the knife and conveyed to the mouth. But this was hardly an easy way to eat. So with the Middle Ages came the advent of the two-knife solution, one knife to steady the food and the other to cut, jab, and convey. Still, as a pointy utensil, a knife does not do a very good job of steadying food on its way to the mouth. For this job, at least two tines are necessary. The fork was born.

Initially, the two-tine fork was used primarily in the kitchen for carving and serving. These were large forks with widely separated tines, and they were useful in this role because the meat could be carved effectively and the fork could be easily removed from the meat. By the seventeenth century, forks began appearing at the dining table in England, but a new problem greeted this transition. Specifically, although a fork with two large, widely separated tines was effective for preventing rotation when carving meat, such a design was not ideal for individual dining because the two large tines were not useful for spearing bite-sized portions of food and, moreover, the wide spacing of the tines was ineffective for scooping. These problems were addressed by the advent of the three-tined fork, which continued to solve the rotation problem. In time, a four-tined fork evolved, followed by brief flirtations with five- and six-tined forks; these flirtations were brief because the width of the mouth sets an effective limit to the width of a fork. Ultimately, the four-tined fork won out, becoming the standard in England by the end of the 1800s.

The fork's impact on the design of the knife has not been trivial. First, the left-hand knife was replaced by the fork as a solution to the rotation problem. Then, the availability of a pointed fork diminished the necessity of a pointed knife. By the end of the seventeenth century, dinner knives were now blunted and, because two-tined forks were the norm at that time and were not ideal for scooping, knives were then given a broad surface to serve a scooping function. Thus has arisen the standard dinner fork and knife. The remaining standard utensils—and the many non-standard ones—have similar histories.

The evolution of eating utensils has shaped the cultural transmission of eating habits in surprising ways. For example, the European style of eating is derived from the history just reviewed, with the knife in the right hand cutting the food and pushing it onto the fork in the left hand for conveyance to the mouth. For some reason, forks were rare in colonial America but spoons were not. Thus, these colonists would use the knife in the European tradition but use the base of the spoon in their left hand to steady the food while cutting, after which the right (and typically more dexterous) hand would lay down the knife and pick up and flip over the spoon to scoop up the food for eating. This crisscross eating method became an ingrained American cultural tradition and therefore survived the introduction of the fork. These European and American behavioral traditions continue today despite the fact that forks, knives, and spoons are placed on the table in identical positions in

Europe and America; in other words, our history with eating utensils is a hidden force that continues to shape our culture and behavior.

Thus, human artifacts are no more the independent creations of any single individual than is the evolutionary invention of sonar in dolphins or echolocation in bats. In Petroski's words, artifacts

> do not spring fully formed from the mind of some maker but, rather, become shaped and reshaped through the (principally negative) experiences of their users within the social, cultural, and technological contexts in which they are embedded. . . . Imagining how the form of things as seemingly simple as eating utensils might have evolved demonstrates the inadequacy of a 'form follows function' argument to serve as a guiding principle for understanding how artifacts have come to look the way they do. . . . If not in tableware, does form follow function in the genesis and development of our more high-tech designs, or is the alliterative phrase just an alluring consonance that lulls the mind to sleep?

None of this is to downplay the significance of the individual's ability to identify and solve problems related to the design of low- and high-tech devices. Petroski does, however, describe the process of invention in a way that is more satisfying than are facile appeals to human consciousness and other indefinable qualities. Humans are clever, ingenious, inventive, resourceful, and persistent. But we do not invent complex, or even simple, devices from a standing start through the mere application of our mind to a problem. We are not that smart.

The argument from design is more than a fallacious argument: it is a reflection of how the human mind works. When we are confronted with complexity and see no path to how that complexity originated, the appeal of the argument from design is immense:

What explains the ability of cats to open gates? *Design, through the action of imitation.*

What explains the ingenuity of human invention? *Design, through the action of reason.*

What explains the complexity of the human eye? *Design, through the action of divine creation.*

Of course, nothing is explained by these answers but, even worse, they numb the very impulse that might lead us to ask deeper and more profound questions. We are lulled to sleep.

Logic is one of the crowning achievements of the human mind, testimony to the promise of human reason. But logic does not guide human thought and action so much as it describes it. Yes, we can apply logic to a problem but, as we know, it takes training and practice to do so effectively. Moreover, we should be cautious about singing the praises of our logical and reasoning abilities when so many of our species' greatest thinkers have declared their allegiance to the argument from design even after its flaws were exposed.

Although my aim here is not to diminish the estimation of our capacity for complex thought, there is little doubt that our intellectual conceits have erected a wall between humans—guided by reason—and animals—guided by instinct. Darwinian thinking, however, does not abide such artificial and arbitrary barriers. So it is no surprise that from Darwin's time to the present day, many who wish to tear down this wall have sought to do so by demonstrating human *instincts* and animal *reason.* Whether one views such attempts as successful or not is related, in no small degree, to one's comfort with such terms as *reason* and *instinct* in the first place.

We have seen in this chapter the pervasiveness of designer thinking and, just as important, we have seen the many benefits to our understanding of the world around us by moving beyond the attractions of designer thinking and asking the next question about origins. When we ask questions about origins, we defeat designer thinking.

Of course, defeating designer thinking when the subject is silverware is one thing. Defeating the appeal to genes as the designers of instinctive behaviors is quite another. And there can be no disputing the fact that—by and large—genes have become biological royalty, imbued with divine powers that are denied all other mere molecules. For those enraptured, the complexities of behavior cannot be comprehended without appealing to a genetic blueprint, a genetic controller, or a genetic program.

Sound familiar?

# 3

# SPOOKY

WITH GREAT FANFARE IN MID-FEBRUARY 2001, the preeminent scientific journals *Science* and *Nature* simultaneously published reports announcing the sequencing of the human genome. One report emerged from a private company, Celera Genomics, and the other from a publicly funded consortium headed by Francis Collins, director of the National Human Genome Research Institute. That the human genome was able to be sequenced at all had been doubted by many, but that the genome was sequenced so quickly amazed many more. Hailed as the greatest triumph of Big Science since the Manhattan Project, the Human Genome Project (HGP) was widely perceived as opening the door to a new universe of self-understanding.

Within only a few months of the *Science* and *Nature* reports, however, the grand pronouncements regarding the implications of the new findings were being replaced with more sober and forward-looking assessments of the next stage of research. Some were already beginning to stress the limited knowledge gained by the sequencing of the human genome. What was needed now, according to this view, was a comprehensive assessment of the actual proteins that the newly identified genes encode. Called *proteomics,* this was to be the next frontier for Big Science.

There were, apparently, even more limitations to be acknowledged. For instance, when I opened my June 25, 2001, issue of *The New Republic,* a magazine that has been required reading in my family for decades, I found an article authored by Francis Collins, the aforementioned director of the National Human Genome Research Institute, and two colleagues, Lowell Weiss and Kathy Hudson, the latter the assistant director of Collins's institute. This article, entitled "Heredity and Humanity," was accompanied by the teaser subtitle: "Have no fear. Genes aren't everything." Breathing a sigh of relief, I read on.

"Unfortunately," according to Collins and his co-authors, "the new focus on the genome has left some people with the impression that DNA's power is perhaps *too* considerable. . . . Fortunately, ten years of intensive study of the human genome have provided ample evidence that these fears of genetic determinism [the notion that genes produce our bodies and our brains according to a fixed plan] are unwarranted." These statements are disingenuous at best. After all, the enormous expense of the HGP was repeatedly justified by its proponents (a group that obviously included Collins) to the American taxpayers precisely on the basis of the enormous benefits to be reaped by harnessing the awesome power of the gene. Moreover, it is misleading to suggest that the "fears of genetic determinism" were only alleviated upon completion of the HGP. On the contrary, the shaky conceptual and logical foundations of genetic determinism have been appreciated for decades, and thus the "discovery" conveyed by Collins and his co-authors is no more groundbreaking than the observation by the first astronauts that our planet is, indeed, not flat.

In fact, in their refutation of genetic determinism, Collins, Weiss, and Hudson barely invoke the findings of the HGP. Rather, they recite a number of arguments that have been around for many years, long before the HGP was even conceived! So why use the announced completion of the HGP as an opportunity to refute genetic determinism? The reason is that the primary aim of the "Heredity and Humanity" article is to ward off the notion, which they fear the HGP will engender, that belief in the powers of DNA is an acceptable substitute for a belief in a "higher power." Thus, the authors argue that those features that represent the core of our humanity—morality, love, and self-sacrifice—can only be understood through belief in God, who is "more than nature." Big Science meets the Big Kahuna.

For Collins and many others, illuminating the fine details of the human genome provides an opportunity to glimpse the mind of the god that created it, a sentiment that is echoed by the popular description of the sequenced human genome as the Book of Life. For the non-religious, however, DNA is often viewed as a creator in its own right, imbued, as historian and philosopher Evelyn Fox Keller has noted, with "a kind of mentality—the ability to plan and delegate." One prominent geneticist, David Baltimore, has even referred to DNA as "the cell's brain." But while the words may differ, the underlying sentiment is the same: for behind these appeals to God, genes, and mind lurks the guiding hand of design.

Ironically, the HGP has shown us that there are, in fact, many fewer human genes than was previously suspected. Consider this: Before the initial report of the HGP was released in February, 2001, it was estimated that the human genome contains between eighty thousand and 140,000 genes. That number was reduced to just thirty-one thousand upon release of the HGP reports and, only two years later, an additional six thousand genes were relieved of duty. The current estimate, then, leaves our species with just twenty-five thousand genes, less than twice that of a fruit fly and even less than that of some plants. Thus, the HGP has unexpectedly increased the burden on non-genetic factors for explaining organic diversity, thereby frustrating the hopes of many that we are on a glide path to enlightenment. As Keller has observed, the HGP has been invaluable "in helping to reveal the naïveté of those

hopes and thereby to set us on a more realistic track toward understanding how organisms develop, function, and evolve . . ." Staying on
this realistic track, however, requires a balanced view of gene function,
one that acknowledges the genome's organismal role without infusing it
with magical powers.

## Magic and MISTRA

Perhaps the most dramatic examples of alleged genetic control over
behavior come from studies of identical twins who were reared in separate homes and reunited as adults. The use of twins to study the
nature-nurture problem is now standard practice and rests on a simple
premise: Given that identical twins result from the equal splitting of a
fertilized egg and therefore share 100 percent of their genes, and given
that fraternal twins result from the fertilization of two separate eggs by
two separate sperm and therefore share (on average) only 50 percent
of their genes, a comparison of identical and fraternal twins provides
an opportunity to compare siblings who are born at the same time and
into the same environment but whose genetic endowment differs by a
known amount.

For the present purposes I will ignore the wrongheaded assumption,
critical to this research, that the environmental influences on identical
and fraternal twins are indistinguishable. I will also ignore the check-
ered past of this research, a past that has been documented by many
others. Rather, my aim here is to show how magical conceptions of how
genes influence mind and behavior make it possible to believe extraordinary things on the basis of the flimsiest of evidence.

Coincidences, especially strange ones, have a way of capturing our
attention and, for some reason, coincidences about twins are particularly
captivating. For example, the story of twin girls, separated at birth in
Mexico and reunited as adults on the campus of Hofstra University, was

recently reported in the *New York Times*. It was a touching story made magical by an "uncanny parallel" in the lives of the twins: each twin's *adoptive* father had died of cancer. The effect on the reader was palpable. Here was a case of identical twins with an eerie, mystical connection that demonstrates the awesome power of the gene. In this case, however, it is indisputable that genes could *not* be involved in the death of an adoptive father. Nonetheless, this story made a splash in part because it fed into a popular perspective regarding twins and genes, a perspective that has been reinforced by many in the scientific community.

One psychologist, Thomas Bouchard, has been at the forefront of the modern study of the genetic basis of personality, largely as leader of a project called the Minnesota Study of Twins Reared Apart (MISTRA). The idea behind MISTRA is that examining the degree of similarity between identical twins who have lived apart for most of their lives allows us to assess the role of genes in determining personality. Using statistical and analytical techniques that are standard fare in their field (despite their well-established shortcomings), Bouchard and his colleagues present their work in the usual scientific fashion in high-profile scientific journals.

The real fame for the MISTRA program, however, has been generated by the widespread dissemination to the public of a series of anecdotes that purport to demonstrate the remarkable power of genes over our personalities. For example, two extraordinarily famous twins, Jim Springer and Jim Lewis, were raised by different families in Ohio, with Jim Springer being raised by his biological parents. After their reunion at the age of thirty-nine, Bouchard and his colleagues noted that each of the "Jim Twins" had been married twice, first to a Linda, then to a Betty; their sons were named James Alan and James Allan; as children, they each had a pet dog named Toy; when they grew up and got married, each twin's family vacationed in the same beach area of Florida and they drove to their vacation spot in light-blue Chevrolets; and both twins smoked Salem cigarettes and occasionally drank Miller Lite beer.

There are many reasons, described by others, to doubt the accuracy and significance of these anecdotes. But let's leave aside these criticisms, assume the anecdotes are true, and assess their implications. Even if true, what are we willing to conclude from these anecdotes about the ability of genes to control behavior? Do we really believe that one gene, or even one thousand genes, can determine the English name of our spouse, child, or dog? Do we really believe that genes can direct us to a vacation spot called Florida, and that they can steer us to buy a particular make of car in a particular shade of blue? Do we really believe that genes can control which brand of cigarettes or beer we prefer?

For many scientists, it seems that the answer to these questions is yes. For example, in *The Language Instinct,* Steven Pinker discusses the MISTRA anecdotes and concludes:

> Many people are skeptical of such anecdotes. Are the parallels just coincidences, the overlap that is inevitable when two biographies are scrutinized in enough detail? Clearly not. Bouchard and his [colleagues] are repeatedly astonished by the spooky similarities they discover in their identical twins reared apart but that never appear in their fraternal twins reared apart. Another pair of identical twins meeting for the first time discovered that they both used Vademecum toothpaste, Canoe shaving lotion, Vitalis hair tonic, and Lucky Strike cigarettes. After the meeting they sent each other identical birthday presents that crossed in the mail. One pair of women habitually wore seven rings. Another pair of men pointed out (correctly) that a wheel bearing in Bouchard's car needed replacing. And quantitative research corroborates the hundreds of anecdotes.

Similarly, Marc Hauser, in *Wild Minds,* relates one of the MISTRA anecdotes and states, "We are awed by these similarities because, of course, the twins were reared apart. So unless we are willing to grant them some form of telepathy, implemented at birth and throughout large portions of their adult life, then we must assign such similarities to their underlying genetic similarity."

These statements are notable for their lack of even a hint of skepticism. Consider Pinker's statement that Bouchard and his colleagues never find "spooky similarities" in fraternal twins reared apart. Never? Think about that. They never find that non-identical twins reared apart prefer the same toothpaste, shaving lotion, hair tonic, and cigarettes, or a similar combination of other products? They never find that non-identical twins drive the same car or have spouses with the same name? Is it really so astonishing that twins (who, of course, share the same birthday) would mail each other birthday gifts, even the same birthday gift, after a visit? But even more important, what are the quantitative analyses that Bouchard and his colleagues perform to corroborate their anecdotes? Have they calculated the probabilities associated with individuals from particular backgrounds using particular products and engaging in particular activities and then, armed with this information, examined the similarities between identical and fraternal twins reared apart? If they have performed such analyses, they have preferred to keep them to themselves.

Consider this college-age pair and the striking similarities between them: "Both are Baptist; volleyball and tennis are their favorite sports; their favorite subjects in school were English and math (and both listed shorthand as their least favorite); both are studying nursing; and both prefer vacations at historical places." These two individuals, however, are not identical twins reared apart, but completely unrelated individuals, matched for age and sex, who took part in a 1984 study. Obviously, "astonishing" similarities are only astonishing when we don't have an accurate account of how common such similarities are between unrelated individuals.

Regardless, you may find the objections to the accounts of the "spooky similarities" between identical twins reared apart to be unreasonable.

Assuming they are true, you might insist (with Pinker, Hauser, and many others) that the similarities really are too striking to dismiss as coincidence. But a failure to demand precise quantitative analyses of the probabilities associated with these anecdotes is a failure to acknowledge just how awful most of us are at assessing probabilistic events in our lives. And it is the objective assessment of probabilities that lies at the heart of scientific inference.

To appreciate this point, consider the following "spooky" fact: In a room filled randomly with only twenty-three people, there is a 50 percent chance that two of them share the same birthday. How is it that 365 possible birthdates can be reduced to such a manageable number? Most people confronted with this problem believe that at least one hundred people are necessary to reach the 50 percent mark. Once incredulous myself, I reviewed the mathematical explanation and have since consistently demonstrated the phenomenon in front of my classes. As a consequence, I no longer consider it spooky.[1]

Or consider the Monty Hall problem, named after the host of the once-popular television game show *Let's Make a Deal!* The problem is based on a central feature of the game and was apparently discovered just before the show went off the air in 1977. It works like this: The contestant sees three doors (#1, #2, and #3), behind only one of which lies the big prize. Say the contestant incorrectly chooses, for example, Door #1. Monty then reveals that the big prize is *not* behind Door #2 and asks the contestant if she would like to stick with Door #1 or switch to Door #3. What should she do? The answer is completely counterintuitive. To improve her odds of winning, she should switch to Door #3. When a popular magazine columnist, Marilyn vos Savant, presented this problem and its solution in a series of magazine articles in the early 1990s, she was met with a barrage of angry letters, including many from professional statisticians who insisted that her solution was wrong. Indeed, the Monty Hall problem appears to be unique in its imperviousness to higher education and academic achievement and in the anger it provokes in those who doubt the correct solution. Regardless, believe it or not, a player of the game increases her odds substantially by switching doors.[2]

The birthday brainteaser and the Monty Hall problem teach us something very deep about ourselves: we are poor intuitive judges of the complex relations that exist among probabilistic events in the world. As a consequence, we should be wary of gut assessments and resist the temptation to leap to fantastic conclusions and conjure miraculous explanations when faced with the uncertain, the improbable, or the ambiguous. So, what are our options in the face of the "spooky similarities" between identical twins reared apart? Hauser would have us believe that there are only two possible explanations—telepathy or genetic similarity—and he evidently supposes that any critical thinker, when faced with this dichotomy, will be repelled by the magic of telepathy and attracted to the science of the gene. But assigning to genes control over our choice of car manufacturer, hair tonic, and spouse's name is as magical and unscientific as telepathy. To dislodge ourselves from the grip of Hauser's false dichotomy, we must clarify what it is that genes actually do.

## *A Messy Pile of Protein*

Long before anyone imagined the existence of genes, scientists struggled to understand the nature of inheritance: How do traits pass from one generation to the next? What is the material basis of heredity? Do our individual life experiences pass to our children, or do our children only inherit that which we inherited? Many wrong turns were made until Gregor Mendel's work on the rules of inheritance—originally conducted in the mid-nineteenth century—was brought to the attention of mainstream biologists in 1900. The flurry of activity that ensued was dizzying. Within the next ten years, the words *gene* and *genetics* had entered our vocabulary. In 1953, the structure of the genetic material was identified as DNA. Soon thereafter, the so-called genetic code was broken. In the 1960s, the genetic revolution joined forces with the computer revolution and the analogy between genes and computer programs was introduced. In 1976, the evolutionary biologist Richard Dawkins took an unconstrained metaphor to its logical extreme in *The Selfish Gene.* In 1997, Dolly the sheep was cloned. And in 2001, the new millennium began with the completion of the first phase of the Human Genome Project.

This furious pace of progress meant that the sober reflections of many individuals on the true functions of genes were not heard. Although some (like Francis Collins) are conveniently arguing for sobriety now that the party has ended, many were doing so just as the party was getting started. These critics were raising arguments, based on logic and scientific fact, that called into question the simplistic notions that were dominating the discussion. One such critic was Gunther Stent, a molecular biologist who was himself an important early participant in the molecular revolution. In response to a half-serious suggestion by Carl Sagan that transmitting the DNA of a cat to aliens on another planet would be tantamount to transmitting the cat itself, Stent argued forcefully that the aliens "would have to know a good deal more about terrestrial life than the formal relations between DNA . . . and protein amino acid sequences." In fact, even if we grant the aliens the ability to use the cat's DNA to produce amino acids and proteins, all they would have (at best) is a messy pile of protein.

To move beyond Sagan's overly optimistic view of the powers of DNA and toward Stent's more balanced and realistic perspective, it is necessary to dispense with some cherished but misleading notions regarding genes and gene expression. One such notion is the "central dogma" of molecular biology, first formulated in 1957, which envisioned a tidy, linear process governing the translation of the genetic code into protein. One gene, one protein—or so we thought. As we now know, the process that governs the translation of the genetic code into protein is so convoluted and context-dependent that not only can a single segment of DNA contribute to the production of many different proteins, there are even proteins that cannot be traced to the code within any single segment of DNA. Evelyn Fox Keller has written that these recent findings, taken together, "threaten to throw the very concept of 'the gene'—either as a unit of structure or as a unit of function—into blatant disarray." To help us appreciate the magnitude of this disarray, she employs the following musical analogy: The "problem is not only that the music inscribed in the score does not exist until it is played, but that the players rewrite the score . . . in their very execution of it."

The implications of this emerging story may seem revolutionary, but this story has roots that go back many years. For even before the historic report of DNA's structure in 1953 by Watson and Crick, it was well known that every cell in the body—brain cell, liver cell, skin cell—contains an identical complement of DNA (which must be the case because all of the cells in our body are derived from a single cell, the fertilized egg). For example, in 1934, nearly twenty years before Watson and Crick's discovery, the geneticist T. H. Morgan wrote that the "implication in most genetic interpretation is that all the genes are acting all the time in the same way. This would leave unexplained why some cells of the embryo develop in one way, some in another, if the genes are the only agents in the results." Same DNA, very different cells. The key, then, to understanding what it is that genes do is to begin to see them as just a part of the machine. Not the brains. Not the brawn. Not even as privileged or exclusive repositories of information.

This last point is the key to a sensible perspective of gene function but is perhaps the hardest for many people to grasp. DNA, as commonly understood, contains information in the form of a code that the cell uses to build amino acids and complex proteins. But although this is where the contribution of the gene ends, the realization of the protein's function is only beginning because proteins do not achieve their function in the one-dimensional, linear world of a genetic code. Rather, proteins must first contort themselves into a complex three-dimensional structure whose final form bears no programmatic relationship to DNA. In other words, the limited value of the information provided by DNA is evident at the earliest stage of gene expression.

Consider a protein like hemoglobin that functions to carry oxygen in our blood. At normal body temperature, hemoglobin exists as a complexly coiled molecule and it is only in this form that hemoglobin has any use. When the temperature of a hemoglobin molecule is raised, however, it uncoils and reverts to a long, flat, and functionless strand (interestingly, we say that the protein denatures). Now, if the hemoglobin is returned to body temperature, it will spontaneously self-organize back into its coiled and fully functioning form. These temperature-induced modifications of the structure and function of hemoglobin occur without any genetic contribution. Moreover, nowhere in the gene that codes for hemoglobin can one find a specification of the optimal temperature for producing the fully functioning form of hemoglobin. In other words, the production of a functioning molecule of hemoglobin is an epigenetic process that depends just as critically upon an obliging environment, including temperature, as it does on DNA.

This is what Stent meant when he wrote that Sagan's aliens "would have to know a good deal more about terrestrial life than the formal relations between DNA . . . and protein amino acid sequences." The environment—not only DNA—is a repository of information that is essential for constructing an organism.

To sum up, at no point in time during the development of any organism do genes control, program, or construct. They certainly never think or design. Indeed, the gene concept itself is losing its grip and, as a consequence, room is being made for a greater appreciation for the roles that the maternal cytoplasm, maternal diet, and other extra-genetic factors play in orchestrating the expression of DNA. In other words, genes are no longer the sole source of cellular control. As will become clear, they aren't even the sole mechanism of inheritance. But we must be careful not to dethrone one despot only to install another. Rather, we need to appreciate that development is an epigenetic process, characterized by nonhierarchical, multicausal, and convoluted pathways that nonetheless effect the stable construction of an organism from one generation to the next. It cannot be overemphasized that proponents of the epigenetic perspective are not advocating mere interactionism—where genetic and environmental causes receive equal emphasis—but a fundamental reworking of how we think about development.

NOTES

[1] Grasping the solution to this puzzle begins by recognizing that we are not seeking the probability of someone else in the room having the same birthday as you (which is how we are used to thinking of such problems), but the probability that any two people in the room will have the same birthday. Imagine a room with one person in it. The probability that her birthday is unique is, obviously, 1. Now add a second person to the room: The probability that this second person's birthday is unique is huge, equal to 1 x (364/365), which means that there is a 99.7 percent chance that the two birthdays are not the same (or just 0.3 percent that the two birthdays are the same). We keep going: Add a third person and we see that the probability that the three birthdays are not the same is 1 x (364/365) x (363/365), which means that there is a 99.2 percent chance that the three birthdays are not the same (and a 0.8 percent chance that at least two of the birthdays are the same). If you continue along with this logic, you find that with twenty-three people in the room, there is a 51 percent chance that two of them will have the same birthday. Moreover, with only sixty people in the room, the odds skyrocket to 99 percent.

[2] Perhaps the easiest way to convince yourself that switching doors is the best strategy is to get three playing cards, set up the game with one card representing the door with the prize, and execute one hundred trials of the game adopting the switch strategy and one hundred trials adopting the no-switch strategy. You will find that, on average, you double your chances of winning when you use the switch strategy (from 33.3 percent to 66.7 percent).

Here, however, is the explanation that I find easiest to grasp intuitively: Imagine that there are one hundred doors (rather than three) and Monty asks you to pick one. You pick Door #57 and you have a 1 percent chance of being correct. Now, Monty proceeds to open door after door until only your original choice and Door #42 remain. What would you do? On one hand, you selected Door #57 at random; on the other hand, Monty opened only those doors that he already knew did not contain the prize. Think of it this way: There was a 99 percent chance that the prize was behind one of those ninety-nine doors, and that enormous likelihood has now been compressed into that single remaining door (#42). The choice is clear.

The same logic applies to the three-door game, it's just that the advantage of the switch strategy is less stark.

# BOUNDARY ISSUES

**WHAT DO WE MEAN WHEN** we say that a behavior is genetic? Imagine a mother and her daughter, both of whom share a talent for music; or a love of swimming; or even an identical high-pitched, explosive laugh. When we see that the daughter shares these traits with her mother but her younger sister does not, we feel justified in concluding that the older daughter alone inherited the mother's music genes, swimming genes, or laughter genes. After all, the two sisters share the same environment, right? Same parents, same house, same relatives, same town, same culture, same religion, same socioeconomic status.

One flaw in this line of reasoning, however, is the assumption that developmental experiences are simply a collection of broad categories—parents, house, town—as if the actual local interactions of the individual with her microenvironment are mere details with little consequence. It is, in actuality, a matter of perspective. From a distance we see a family, culture, and money, but as we zoom in we see, for example, that the younger sister's environment is occupied by her older sister (who once commanded her parents' undivided attention), that her parents are older and more experienced than when the older sister was born, that they are now making a better living but have less time to spend with the kids. Consequently, the older sister must often look after the younger sister when their parents can't spare the time, an arrangement that often leads to fighting and resentment. Development, in other words, is a turbulent affair that is not realistically captured by check-off boxes on a questionnaire.

It is unfortunate that the mere act of engaging in the above discussion further fortifies the entrenched notion that genetic and environmental factors act independently to shape behavior. As we will see, they do not. But even if we were to insist that the dichotomy is valid, it isn't clear what we would ever gain by saying that a behavior is genetic. On one hand, we could say that every behavior is genetically determined in that some genes must help to produce it; on the other hand, we could say that no behavior is genetically determined in the sense that genes alone are never sufficient to produce it. In other words, referring to a behavior as genetic can explain everything or nothing depending on how we use it, making it a rather empty concept.

Similar confusion bedevils a related and widely used term: heritability. The confusion arises first because many of us don't realize that heritability is not simply another word for inherited. Rather, heritability is a technical term that expresses a relationship between trait variability and genetic variability. In trying to understand what heritability means, the word *variability* is key. Consider that only 5 percent of all human genes differ between individuals (that is, are *polymorphic*); this means that 95 percent of all human genes are identical in every member of our species. Obviously, any trait that arises from one of those 95 percent of genes would qualify as being genetic by most definitions of that term. And yet, because those genes do not vary, the concept of heritability cannot apply to them. Indeed, in a world filled with clones, no trait is heritable. No variability, no heritability.

Still, the 5 percent of genes that are polymorphic help form the basis for those traits that distinguish one individual from another. Even here, however, the usefulness of a heritability analysis is more limited than most people think. For example, behavior geneticists Dean Hamer and Peter Copeland write that "the heritability of male sexual orientation is about 50 percent. That means that being gay is about 50 percent genetic and 50 percent from other influences. . . ." Is this what heritability really means? And does this imply that there exists a "gay gene" that, when present, produces a gay individual 50 percent of the time? To convince you that the answer to both questions is "no," let's consider a trait with an even higher heritability than sexual orientation.

In humans, height is a highly variable trait with a heritability close to 90 percent. Tall parents tend to have tall children and short parents tend to have short children. Does this mean that "being tall is about 90 percent genetic and 10 percent from other influences?" Anyone who has stooped low to pass through the doorways of houses only several hundred years old—as I remember doing when I visited Stratford-upon-Avon, the birthplace of William Shakespeare—recognizes how changes in the environment can impact growth rates in our species. Human genes haven't changed substantially since Shakespeare's time, but nutrition has.

Furthermore, a substantial heritability for height only means that the genetic variability of parents is associated with height differences of offspring *for a specific group of individuals living in particular environmental conditions.* Thus, if you transport a group of adults—tall and short—to a new locale with different food, climate, and standard of living, you may find that the heights of the offspring are no longer closely related to the heights of the parents. In other words, heritability is not fixed.

Having said all this, you may be disappointed to learn that even when used and interpreted correctly, heritability analyses are plagued by a fundamental flaw—a flaw that rests on the assumption that genes and environment make independent and separable contributions to development. Geneticists Richard Levins and Richard

Lewontin use a construction metaphor to illustrate the problem with this assumption:

> [I]f two men lay bricks to build a wall, we may quite fairly measure their contributions by counting the number laid by each; but if one mixes the mortar and the other lays the bricks, it would be absurd to measure their relative quantitative contributions by measuring the volumes of bricks and of mortar. It is obviously even more absurd to say what proportion of a plant's height is owed to the fertilizer it received and what proportion to the water, or to ascribe so many inches of a man's height to his genes and so many to his environment.

Other writers have devised similarly powerful metaphors. For example, we compute the area of a rectangle by multiplying length by height, but we would never ask: What contributes more to a rectangle's area, length or height? And Simon Blackburn writes that "it is meaningless to ask whether iron rusts because of the nature of iron or because of the environment in which iron is put." Nevertheless, the confusion surrounding heritability lives on, as those whose scientific careers have been built on heritability analyses stubbornly cling to the assumption that genetic and environmental contributions to development are separable.

Metaphors are useful pedagogical tools but no one is convinced by them alone. Ultimately, metaphors must be backed up with facts. That is the aim of this chapter. And as we will see, the line of demarcation between genome and environment that exists at conception blurs and eventually disappears with each passing moment of developmental time.

### *Location, Location, Location*

Perhaps the most convincing way to illustrate the false dichotomy between genetic and environmental causes is by demonstrating the impossibility of identifying a clear boundary between them. It is important to stress that such demonstrations, including those discussed below, are not exceptions to a rule. Rather, they reflect the true nature of development as a bidirectional, dynamic, and often nonobvious cascade of events that defies simple partitioning into straightforward causes.

At the earliest stage of development in chordates (a group that includes fish, amphibians, reptiles, birds, and mammals), the unfertilized egg is a single-celled sphere that consists of genetic material from the mother and the cytoplasm—the liquid substance that contains the yolk and other ingredients. It is here, within the egg, that we can discern the two developmental perspectives—preformationism and epigenesis—that have shaped embryological disputes for centuries. The embryologist Jack Cohen described the modern rendition of this dispute:

> Two diametrically opposed views of egg function were current in the 1950s and 1960s. One saw the egg [as a passive entity] which functioned by remaining relatively immobile while awaiting discovery and penetration by the sperm. . . . The other viewed the egg as a complex, balanced machinery awaiting activation by the sperm, a kind of 'Sleeping Beauty' model. By and large, geneticists and DNA-is-God-and-RNA-is-his-prophet molecular biologists took the first approach, and embryologists saw the egg as a beautiful but willful maiden awaiting the Prince's kiss.

Flowery language aside, the critical point here is that the egg's cytoplasm does not exist merely to bathe the genome-containing nucleus, but rather "is best considered as a *structured ambience* inherited by each embryo in addition to its own genome." This notion of a structured ambience is at the heart of the concept of epigenesis: like a hand and glove, the environment provides a structure that complements the genome.

The first critical events after fertilization—including when the single-celled egg first cleaves into two cells—proceed without any guidance from the embryo's own genes. They are completely inert. Indeed, in some animals, thousands of cell divisions can occur before the first sign of embryonic gene expression is detected. At these earliest stages of this critical process, the forces that structure the ambience within the egg can be so mundane as to be easily overlooked. For example, under the influence of gravity, the heavier constituents of the cytoplasm settle in the lower half of the fertilized egg, just as sediment settles at the bottom of an upright wine bottle. This creates a nonuniform distribution of materials that sets the stage for all succeeding developmental events. In other words, for many animals, including chickens and humans, gravity literally grounds the process of normal development to our planet; as a consequence, attempts to gestate these animals in the microgravity of space are likely to fail.

Once embryonic gene expression commences, local cytoplasmic factors continue to exert their influence, modulating which genes will be expressed at which times. Guided by these cytoplasmic microenvironments, it is the local activation of different genes within the zygote that brings about the development of the three primary tissue layers that correspond to skin and nervous system (ectoderm), muscle and skeleton (mesoderm), and internal gut organs (endoderm). Tip O'Neill, the former Speaker of the U.S. House of Representatives, famously observed that all politics is local; the same is true for gene expression. This is one reason why Gunther Stent's aliens, staring at the DNA of a cat, would scratch their heads in wonder. Genes alone simply do not tell the whole story.

Scientific insights are often gained through the intensive study of a species that exhibits particularly odd but useful characteristics. For the study of embryogenesis and gene expression, the fruit fly, *Drosophila*, has proven its worth. While the embryos of many animals (as described above) divide completely and repeatedly during the early developmental stages, in flies, only the nuclei (which contain the genetic material) divide, producing a hollow sausage-shaped and cytoplasm-filled single-celled embryo with nuclei distributed along the outer edge, like race fans at the Indianapolis 500. The action begins in the front of the embryo with the production of a protein, called Bicoid, that spreads down the length of the embryo like an ink drop at one end of a lap pool, producing a chemical gradient that influences the expression of other genes and, therefore, subsequent development of the embryo. For example, if the concentration of Bicoid is increased so that its influence spreads further down the length of the embryo, then the fly may develop an enlarged head but no thorax (the middle section of a fly that contains its heart and to which its wings are attached). Most freakishly, if the gene that codes for Bicoid is transplanted to the posterior end of the embryo, then the fly will develop a second head where the abdomen would normally reside.

Thus, not only is Bicoid protein the product of genetic activity, but it feeds back on the system to influence other genes and other gene products. This is a critical point that cannot be overemphasized because such looping cascades of influence make laughable any attempt to divide an organism into genetic and environmental causes. For how does one properly characterize the influence of Bicoid in the fly? Is Bicoid an environmental stimulus or, by virtue of its origins, a genetic stimulus? What percentage of the fly's head is genetic and what percentage is environmental? As the behavioral geneticist Douglas Wahlsten has noted, once "we know how the system of causes works, assigning percentages to the different parts becomes an empty exercise."

As might be expected, the boundary between genetic and environmental causes becomes even more fuzzy as we ascend to higher levels of complexity. Consider phenylketonuria (PKU), a classic example of a grim "genetic disease" that can nonetheless be avoided through a simple

environmental adjustment. The gene in question codes for phenylalanine hydroxylase (PAH), a liver enzyme whose function is to degrade
the amino acid phenylalanine (the primary ingredient of the artificial
sweetener NutraSweet). Infants born with two defective copies of the
PAH gene experience accumulating levels of phenylalanine that, in
turn, result in reduced transport of other amino acids to the brain.
Starved of amino acids and, ultimately, of proteins, the brain fails to
develop normally, resulting in mental retardation. This apparently preordained sequence of events can be short-circuited, however, merely by
preventing PKU infants from eating foods that contain phenylalanine.
Moreover, once brain development is complete, diets containing phenylalanine can be reintroduced without harmful effect. On the basis of
these observations, the molecular biologist Michel Morange has wondered if we would be wiser to designate this prototypical genetic disease
as an environmental disease instead because it is only expressed when
phenylalanine is available in the diet. Indeed, from this perspective, one
wonders how many other PKU-like diseases currently exist without our
knowledge because we have not yet been exposed to the necessary environmental conditions. Perhaps some of these diseases will reveal themselves in the future to our descendants as they encounter radically
different environments on newly colonized planets.

Recent work shows that there is even more to be learned from PKU.
Specifically, consider a woman with PKU who, as an infant, was successfully treated with a low-phenylalanine diet but who now, as an adult,
consumes a normal diet. Because she still lacks the PAH gene and
therefore cannot degrade the phenylalanine that she consumes, she
exhibits soaring blood levels of this now-harmless amino acid. Harmless, that is, for her. Why? Because now she is pregnant, and her normal
non-PKU fetus is now being exposed in utero to levels of phenylalanine
similar to those experienced by PKU infants! Thus, the genes are different but the result is the same: mental retardation caused by over-
exposure to phenylalanine. So how do we assign causation in each case?

Doug Wahlsten frames the conundrum like this: Is PKU a genetic disease that retards brain development via an environmental insult or an environmental insult that arises from a genetic defect in the mother? The line continues to blur.

Now let's consider a biological feature that is so unequivocally "genetic" that entire chromosomes are named after it: sex. Nearly everyone has heard of the male Y chromosome and the female X chromosome. XY means male; XX means female. Case closed.

In fact, there are multiple dimensions in which this simple story is inaccurate. We begin by describing what it is that sex chromosomes actually accomplish. Specifically, we all begin life in utero with an "indifferent" gonad, a term that refers not to the gonad's attitude but to its double identity; at this stage the gonad is equally capable of becoming a testis or an ovary. Which path is taken depends on the presence of an enzyme that, when present, causes the indifferent gonad to become a testis; absence of the enzyme results in production of an ovary. Once this decision is made, many of the differences between males and females—from reproductive organs to brain development—follow like falling dominos. Thus, the presence or absence of the gonad-determining enzyme is a critical early factor and, no surprise, this enzyme is produced with the help of a gene on the Y chromosome.

The surprise comes when we confront a disturbing reality: many animals, including all crocodilians studied thus far and some turtle and lizard species, completely lack sex chromosomes. But, of course, all of these animals have distinguishable sexes like the rest of us. Thus, it is possible for two turtles to be clones, that is, genetically identical in every way, except one has become a male and one a female. To do this, they must have arrived at their sexual destination by taking a non-genetic route. But how?

Remember, one of the early key decisions in human sex determination depends on the production of an enzyme associated with the presence of the Y chromosome. The same is true for turtles, except that production of the enzyme is triggered by the temperature at which the egg was incubated rather than the presence of a gene. This phenomenon—called temperature-dependent sex determination—is expressed in a variety of ways. For example, in some species, warm incubation temperatures produce males and cool incubation temperatures produce females. In other species, the reverse is true. In still others, males are produced at intermediate temperatures. But regardless of the exact pattern, the lesson is the same: The distinction between genetic and environmental causes is a chimera. In humans, production of the gonad-determining enzyme is linked to a genetic difference between the sexes but, in turtles, production of the gonad-determining enzyme is linked to an environmental difference. On what sound logical basis, then, can we say that genetic causes are more privileged than environmental causes?

Let's consider another example, this time further along the developmental path. Many species of mammals, including dogs, cats, mice, and gerbils, produce many offspring within a single litter. These litters are composed of young of both sexes. Imagine two adult female mice, sisters that were fathered by the same male, raised together in a burrow with their mother and now living near each other in an open field. We notice a remarkable difference between these two siblings: One is considerably more aggressive, holds and protects a larger territory, exhibits higher levels of testosterone and lower levels of estrogen, reaches puberty later, is less attractive to males, and even on occasion mounts other females in a way that we typically associate with males. It is as if one of the sisters is less typically feminine than the other and, in some ways, more masculine. But how could two siblings of the same sex, with the same father, same uterine environment, and same rearing environment display such radically different behaviors as adults? Certainly we would be on the right track if we hypothesized that the sisters had inherited different genes from the parents; you might even wonder if the male-like female had inherited a "masculinizing gene." Indeed, if you were told that the male-like female tended to give birth to female offspring with similar male-like traits, you would no doubt be convinced that a "masculinizing gene" was involved. But alas, it isn't so.

A clue to the cause of this sibling difference comes when we examine closely the genitalia of newborn mice. This is not mere voyeurism. What we are looking for in particular is the anogenital distance, that is, the distance between the genitalia (which are tiny bumps on the abdomen in newborns) and the anus. The anogenital distance in males is greater than that in females and, moreover, the anogenital distance in females is directly related to the amount of testosterone to which they were exposed during gestation. But here is the puzzle: We know that males produce their own testosterone, thus explaining why their anogenital distances are greater than those of females. Indeed, male mice exhibit surges in their levels of testosterone twice during development, once during a ten-day period around the time of birth and again much later at the onset of puberty. Perhaps the differences seen in the females at birth result from their being exposed to different levels of testosterone as fetuses. But why would females be exposed to high levels of testosterone around the time of birth, and why would some females be exposed more than others?

The answer arrives when we examine how fetal mice are arranged in the uterus. Mice, like many rodents, have two uterine compartments, and the fetuses are arranged in these compartments like strings of pearls. Each fetus lives in a fluid-filled chamber separated from its neighbors by fetal membranes. The separation, however, is not absolute; the amniotic fluid surrounding each fetus can diffuse from one chamber to the other. Significantly, the testosterone produced by a male fetus can diffuse via the amniotic fluid to a neighboring fetus. Moreover, by chance, a female fetus might find herself sandwiched between two male fetuses (designated a 2M female) such that she is exposed to significantly higher levels of testosterone than a female sandwiched between two female fetuses (designated a 0M fetus). It is the 2M female that, under the influence of her local testosterone-filled environment, exhibits the male-like anatomical, physiological, and behavioral features described above.

But we are not done. Under normal conditions, rodents such as mice and gerbils give birth to equal numbers of males and females. If a mother gestates a disproportionate number of male embryos, however, the proportion of 2M females necessarily increases. Interestingly, 2M females produce litters with more males than females, and 0M females produce litters with more females than males; consequently, 2M females produce more 2M daughters and 0M females produce more 0M daughters, thereby passing their gestational environment to the next generation. This is a striking example of the non-genetic inheritance of a robust and biologically meaningful trait.

In the shadow of a century of intense focus on the gene, it is easy to forget that Darwin formulated his theory of natural selection without any understanding of the mechanisms of inheritance. In fact, his theory required only that there be inheritance *of some kind*—that offspring resemble their parents more than they resemble cousins, distant relatives, and strangers. Whether there were one, two, or ten mechanisms of inheritance was of no consequence for the core Darwinian claim that evolution occurs through natural selection. It was only long after Darwin's death that inheritance became inextricably linked with genetics under the rubric of the so-called modern synthesis. Today, however, investigations of non-genetic modes of inheritance are attracting renewed interest. For many weaned on mainstream evolutionary perspectives (like me), the notion of non-genetic modes of inheritance is initially disorienting. But if the dichotomy between genetic and environmental causes really is false, then we should anticipate a continuing, steady accumulation of evidence documenting the fact that we inherit more from our parents than their genes.


Today, the intoxicating influence of simplistic genetic thinking is fading—and a newfound appreciation of epigenetic factors is growing—

as geneticists confront the need to attain a deeper understanding of the regulatory dynamics of the cell. Ironically, one of the catalysts of this new-found appreciation of epigenesis is the very technique—cloning— that many thought would prove the causal primacy of the gene.

Cloning relies on successfully reversing the interwoven connections that emerge during development between the DNA-containing cell nucleus and the structured ambience of the cytoplasm. It is almost as difficult as unbaking a cake. It should come as no surprise, then, that cloning is akin to any process of human invention, replete with tinkering and trial and error. Thus, even those who cloned Dolly the sheep in 1997 do not yet know why they were successful.

Clues to why cloning is so difficult come from recent research into a process called genetic imprinting. It was once thought that each pair of genes—one from the mother and one from the father—turns on and off simultaneously. It now appears, however, that genetic imprinting can cause one member of a pair of genes to be silenced while the other is expressed. Even more surprising, whether a gene was donated by the mother or the father can determine whether it will be expressed.

What mechanism could possibly keep track of which parent donated a gene? To answer this question, recent research has focused on a process called methylation, in which a molecule called a methyl group binds selectively to a segment of DNA and thereby silences it. The pattern of methylation in the genome is inherited separately from each parent but is not encoded using DNA (thus leading some, unfortunately, to refer to an epigenetic program). Even more curious is the fact that the methylation pattern is wiped clean soon after conception but reemerges intact during gestation (a process referred to as epigenetic reprogramming). It is not known how or where this epigenetic information is stored in the cell, although segments of DNA that do not code for proteins—currently called "junk" DNA—may play a role. What is apparent, however, is that successful cloning requires duplication of the genes *and* the methylation pattern. Thus, until the formation of methylation patterns is better understood, cloning will necessarily be a hit-or-miss affair.

Methyl groups are derived from foods, like beets, that are rich in folic acid, vitamin B12, choline, and betain. Thus, given the role of methylation in gene expression as described above, the availability of methyl groups in the diet of a pregnant female should modulate the gene expression of her offspring. This was shown recently in a strain of mice whose coat color varies from yellow to brown depending on whether a specific gene is methylated during development. When pregnant yellow mice were fed a normal diet, the majority of their offspring developed yellow coats. But when pregnant mice were fed methyl-rich diets, the majority of their offspring developed dark coats. Moreover, the coat color of the offspring persisted into adulthood and was even inherited by their offspring. In other words, a simple manipulation of the mother's diet can produce inherited changes in her grandchildren without changing the genes themselves.

The promise of cloning and thus the fear, too, is its ability to produce duplicate animals: farm animals, pets, perhaps even children. But lost in the parade of cloned sheep, mice, cows, goats, horses, and pigs that are attracting so much attention is the astounding observation that clones simply are not duplicates. This point was made most clearly when the first cloned cat, Cc (for Carbon copy), was born on December 22, 2001. The project that produced Cc was supported financially by a dog owner who wanted to duplicate a beloved and departed collie and hoped that cloning would do the trick. It was immediately apparent, however, that Cc, with her gray stripes over a white coat, didn't look anything like her mom, Rainbow, a calico with the typical smattering of brown, tan, gold, and white. And as Cc grew older, many more differences emerged: Rainbow was beefier and more reserved than her lean, playful daughter, for example. Thus, here was direct evidence that—lo and behold—genes alone do not make the pet.

### *Taking Your Licks*

One of the most tantalizing demonstrations of non-genetic inheritance is the culmination of a circuitous, surprising, and often weird series of studies in rats. The story I will tell hinges critically on a phenomenon known as anogenital licking, in which a mother licks the highly localized region between the pup's hind legs that includes the anus and genitalia. I am not kidding. But to understand how anything exciting could arise from studying such a bizarre behavior, some background will be helpful.

Psychobiology is the study of the biological bases of behavior. Developmental psychobiology—not to be confused with developmental psychology—examines this topic from a developmental perspective. This perspective involves more than simply studying infants; rather, the goal is to understand developmental process, the actual paths taken by animals as they grow and change through the various stages of prenatal and postnatal life.

Developmental psychobiology is a well-defined field with its own society and a meeting where a few hundred of us gather each year to share our most recent work. Because it is a relatively small society, many of its members are the intellectual descendants of a small group of individuals, most of whom viewed development from a common perspective. To many, the godfather of the field is Theodore C. Schneirla, a Swiss-born entomologist and comparative psychologist who, as curator of the Department of Animal Behavior at the American Museum of Natural History in New York Ci ty, trained and influenced many important scientists until his death in 1968. One of Schneirla's most accomplished students was Daniel Lehrman, who was the founder of the Institute of Animal Behavior at Rutgers University. Lehrman is still revered among developmental psychobiologists, not only for his groundbreaking work on the hormonal control of behavior in ring doves—my own graduate advisor loved to recount stories of Lehrman, a large man imitating the coos and wing flaps of mating ring doves for his students—but also for his vociferous yet principled attack on the founder of the field of ethology, Konrad Lorenz, in 1953. At the time, Lehrman was only thirty-four years old.

Beginning in the 1930s, Lorenz initiated a program of research that ultimately elevated the study of animal behavior to a level of rigor comparable to that of other natural sciences. Central to Lorenz's approach to animal behavior was the concept of instinct, defined as a pattern of behavior that is rigidly coordinated and reproducible and that, in its totality, is an inherited feature of a species. For Lorenz (and others, most notably Niko Tinbergen, who shared the Nobel Prize with Lorenz and Karl von Frisch in 1973), innateness refers to the notion that instincts are expressed fully formed without prior experience, even in animals that are raised in isolation and that have been prevented from practicing the behavior.

Lehrman argued forcefully that Lorenz's notions of instinct and innateness were based on a misguided conception of how behaviors actually develop. Whereas Lorenz viewed the expression of species-typical behavior by animals raised in isolation as overwhelming evidence in support of instinct and innateness, Lehrman demurred by pointing out that isolated animals have not necessarily been isolated from the critical experiential factors that govern behavioral development. Incisively, Lehrman emphasized that the "important question is not 'Is the animal isolated?' but '*From what* is the animal isolated?'" If you believe that instincts exist in miniature form at birth and reach their adult form through a simple, obvious process of growth and maturation, then you will find isolation experiments compelling. If, on the other hand, you believe that these species-typical behaviors emerge in real time as a result of complex, nonobvious interactions between the organism and the environment that can begin even before birth, then you will look to other kinds of experiments for insight.

At the heart of the dispute between Lehrman and Lorenz was the meaning of experience. For Lorenz, experience was equivalent to learning (a view that today has been adopted by nativists in the field of developmental psychology, as we will see in Chapter 7) and thus isolation experiments, by preventing learning, are sufficient to demonstrate innateness. In other words, Lorenz concluded that an instinctive behavior is innate when it could be demonstrated that there had been no prior opportunity to learn it. Lehrman's view of experience, however, was adopted from his mentor, Schneirla, who argued that experience includes every kind of interaction between organism and environment that contributes to development, not just the formalized forms of learning made famous by Pavlov, Thorndike, Watson, and Skinner. Thus, for Schneirla and Lehrman but not for Lorenz, experience would include the influences of the cytoplasm on gene expression in the fertilized egg, the concentration gradient of the Bicoid protein in flies, incubation temperature in turtles, and intrauterine position in mice. It is no wonder, then, that Lehrman held firm to his belief that labeling a behavior as *innate* "adds nothing to an understanding of the developmental process involved."

Lehrman's attack on Lorenz is filled with so much passion that one cannot help but sense his disdain for the elder scientist. Some of Lehrman's passion can be traced to his military training during World War II when he became fluent in German and, unlike his American and British contemporaries, familiarized himself with Lorenz's lesser-known writings regarding human interbreeding and the future of our species. But Lehrman's disagreement with Lorenz was not predominantly about settling a personal score: Lehrman disagreed just as vigorously with the Dutchman Tinbergen who, although a student and close associate of Lorenz's, eventually befriended Lehrman and acknowledged the validity of his critique. Jay Rosenblatt, another prominent student of Schneirla's, who was a colleague of Lehrman's at the Institute of Animal Behavior, wrote upon Lehrman's death in 1972 that the disagreement between Lehrman and Lorenz was ideological in the sense that it derived from different scientific cultural traditions: "These traditions dictated that for Lorenz the categories innate and learned were sufficient to deal with how behavior became adapted to natural environments during evolution. But [Lehrman] believed these categories were too restrictive and narrow; he required the fluidity allowed by the developmental analysis of animal behavior . . . to understand the variety and complexity of adult behavior."

In Lehrman's influential 1953 paper, maternal behavior in rats was discussed as an instructive example of a behavior—or more accurately a suite of behaviors—that develops in a way that cannot be assigned to either of the dichotomous categories of learning or maturation (the latter being a term used by Lorenz to denote the opposite of learning). Lehrman rejected this false dichotomy for maternal behavior which, even fifty years ago, was already the focus of much psychobiological research. The heyday of the study of maternal behavior was, however, just beginning, and the research that was to follow in the next few decades advanced our understanding of the actions of hormones, neural development, and stress.

The care and feeding of a litter of newborn rats is no small task. As each infant emerges from the birth canal, the mother peels away the fetal membranes, eats the placenta, and gathers her young in a nest. The young find their way to her teats following an odor trail that the mother lays down by licking her own abdomen with amniotic fluid. With all pups gathered under her and attached to a nipple, she hovers over the pups while they suckle; the stimulation provided by the sucking pups is necessary to trigger a reflex that releases milk to the pups. Conversely, pups cannot urinate or defecate without tactile stimulation provided by the mother; this tactile stimulation comes in the form of anogenital licking, which the mother accomplishes by using her mouth or forepaws to flip the pup into a supine position against her warm body, thereby relaxing her pup and giving her easy access to the pup's anogenital region. She then commences a bout of licking which, critically, evokes reflexive urination and, during longer licking bouts, defecation. Mothers repeat this behavior with each of her pups many times throughout the day, beginning at birth and continuing until her pups are weaned.

Although it may appear that pups reap all of the benefits in their interactions with their mother, there is more here than meets the eye. Mother rats not only lick their pups, they ingest what the pups release. Ingesting urine and feces may be disgusting to us, but it is of tremendous value to rats. First, with pups restricted to the nest and with mothers raising their young on their own, ingesting urine and feces allows her to keep a clean house. Second, lactation is an energetically expensive process that increases each mother's need to forage for food and water, thus requiring her to take leave of her defenseless pups and exposing her to the dangers that loom in the outside world. Viewed in this light, anogenital licking is beneficial to the mother because it allows her—safely and cheaply—to reclaim substantial amounts of water and nutrients that she had previously delivered to her pups in the form of milk. In other words, rats recycle.

One of Lehrman's students, Celia Moore, took our understanding of anogenital licking to another level when she observed that, within the same litter, male pups are licked more often than female pups. One question immediately comes to mind: How does the mother know which of her infants are sons and which are daughters? Moore and others found that the mother's focus on male offspring does not derive from knowledge but, at least in part, from a chemical attraction: male pups produce a testicular-dependent odor in their urine that mothers, for their own reasons, find highly attractive. In fact, when male urine is placed on the head of a female pup, copious *head*-licking ensues. But is the mother's attraction to her infant males' urine a simple matter of taste or is there something more going on here?

There is, in fact, a lot more going on. Let's focus on the object of the mother's attention—the genitalia. At the base of the penis is a muscle, the bulbocavernosus (BC) muscle, that is important for penile erection. Infant males and females possess this muscle at the time of birth but it soon disappears in females. Like all such muscles, the BC is composed of many muscle fibers, and each fiber receives stimulation from a nerve that meanders from the base of the penis to the spinal cord. Within the spinal cord, the cell bodies of neurons cluster in a small group that is called the spinal nucleus of the bulbocavernosus (SNBC). (It is an unfortunate happenstance of nomenclature that *nucleus* is used to describe a dense collection of neurons as well as the compartment within cells that contains the genetic material). When an adult male rat is sexually aroused, the brain sends signals down the spinal cord to activate the neurons in the SNBC, which then send signals down the nerve to the BC muscle, producing penile erection.

The SNBC of adult male rats contains approximately two hundred neurons, the same number found at birth in the SNBC of both males and females. What this means is that development of this system entails the loss of these neurons in females, not their accumulation in males. The overproduction and subsequent death of neurons is a basic feature of the developing nervous system and the SNBC happens to be one of the finest examples of this phenomenon. So what factor or factors determines that the male retains his BC muscle and SNBC neurons, while females do not? One part of the answer is that males produce testosterone which, as a steroid, promotes retention of the BC muscle (a characteristic of steroids that inspires their use among body builders). Thus, testosterone helps to retain the BC muscle, which in turn means that the SNBC neurons maintain their connections with that muscle, which in turn provides a signal to the neurons that prevents their death.

A second part of the answer, however, brings us back to anogenital licking. By licking the male pup's genitalia, the mother provides sensory stimulation to the SNBC neurons in the spinal cord. Moore demonstrated the importance of this sensory stimulation by showing that mothers deprived of their sense of smell (and thus prevented from detecting the presence of the attractant in the male pups' urine) no longer licked males more often than females, and males that were underlicked retained fewer SNBC neurons. When Moore artificially stroked the genitalia of these underlicked males, thereby mimicking the stimulation provided by the mother, normal development was reinstated.

Even though Moore demonstrated that anogenital licking increases retention of SNBC neurons in infants, it remained to be shown that licking was important for adult male reproductive behavior. Moore demonstrated its importance by manipulating the mother's propensity to lick her male young and thus creating adult male offspring that had received differing amounts of anogenital stimulation during infancy. When allowed to mate with females, underlicked males were able to copulate, but they did so at a slower pace than normal males; specifically, they appeared to have trouble erecting their penis, causing them to mount females less successfully and thus slowing the entire copulatory process. This finding was particularly important because it provided an important link between early anogenital stimulation and a functionally important adult behavior.

The next episode in the anogenital licking story concerns the inheritance of the proclivity of mothers to engage in anogenital licking. One way to approach this question of inheritance is to examine the natural variability of maternal behaviors within a population of rats and look for associations between the behaviors of mothers and their daughters. Indeed, when this was done, it was found that some mothers reliably exhibited higher rates of licking, grooming, and other maternal behaviors than other mothers. These variations in maternal attentiveness even appeared to be an inherited trait: when mothers provided high levels of licking to their daughters, the daughters grew up to similarly lick their young at high levels. In addition, male and female offspring of these high-attention mothers exhibited reduced levels of stress and fear as adults. In other words, young that received high maternal attentiveness grew up to be less fearful than young that received low maternal attentiveness, suggesting that fearfulness is also inherited. But what is being inherited? Is there a maternal attentiveness gene, a fear gene, or a common gene for both traits?

These questions were answered by cross-fostering pups, a procedure in which pups are transferred to different mothers soon after birth. When Michael Meaney and his colleagues cross-fostered the young of high-attentiveness mothers to low-attentiveness mothers, and vice versa, they found that the traits of the young reflected the traits of their foster, not their biological, mothers. For example, female young provided with foster care by high-attentiveness mothers exhibited less stress-reactivity than their biological siblings cared for by their own low-attentiveness mothers. Amazingly, these foster-raised pups grew up to be high-attentiveness mothers themselves, a stunning example of non-genetic transmission of behavioral traits that is mediated by the mother's behavior toward her young, not by her genes.

Meaney and his colleagues have taken their findings further by showing how differences in maternal care are translated into differences in gene expression in the brain. Previously, they had shown that differences in maternal care during infancy help to establish the sensitivity of the stress response system in adulthood. For example, pups exposed to mothers who engage in high amounts of licking and grooming showed reduced hormonal responses to threatening situations as well as changes in the brain that indicated a dampening of the brain's response to stress. One area of the brain examined by these investigators is the hippocampus, a structure that receives a lot of attention from researchers because of its prominent role in learning and memory, including the ability of animals to learn about and remember the spatial environment.

Taking advantage of current knowledge of hippocampal function, Meaney and his colleagues showed that increased maternal licking and grooming during early infancy alters gene expression within the hippocampus that, in turn, promotes the development of more connections among neurons within this brain structure. As a consequence, when these pups become adults, their spatial learning abilities are enhanced.

A recent startling twist in this story is beginning to resolve the mystery of how maternal style can be inherited by successive generations via a non-genetic mechanism. This resolution brings us back to an issue we encountered earlier in the chapter: methylation. Recall that methylation patterns are wiped clean at conception but reemerge later in development. In rats, it turns out, the methylation pattern is highly modifiable soon after birth and the pattern that emerges depends on the care received from the mother, including anogenital licking. This is an emerging area of research that promises to revolutionize our understanding of the role of experience-dependent gene expression for the developing organism.

I have taken you on this journey through the delightful world of anogenital licking to make a simple but important point. Imagine how impoverished our view of maternal behavior would be had we simply labeled female rats as possessing different personality traits based upon their proclivity to engage in maternal care—close vs. distant, warm vs. cold, accepting vs. rejecting—and then initiated a search for the "maternal instinct gene." How easy this would have been! This is, in fact, how research is often conducted today. Behavior, however, is much too complex to expect that such crude approaches could generate meaningful insights. As we have seen, the path from anogenital licking to adult behavior to inheritance meanders in startling and unexpected ways. The lesson is clear: when confronted with complexity of this magnitude, real scientific progress results when researchers keep an open mind, value rigorous experimentation, understand that naming is not the same as explaining, and resist the quick and easy answer.

## *Too Many Removes*

The road to behavior must pass through the brain, the most metaboli-cally active organ of the human body. Comprising billions of neurons and trillions of connections, the human brain nonetheless develops in a remarkably predictable way from individual to individual. If you take a section of brain tissue, pin it to the wall, and throw a dart at it, chances are that any competent neurologist can give you a general idea of what the impaled part of the brain is for, to what it is connected, and what neurochemicals are produced there. As with any machine of such immense complexity, perhaps even more than any man-made machine, the brain looks well-designed. So when geneticists began turning their attention to behavior in the 1960s, they naturally assumed that genes could fill the role of designer. As Seymour Benzer, one prominent geneticist, wrote in 1971, the "genes contain the information for the cir-cuit diagram" of the nervous system.

Although Benzer's statement was naïve in 1971, it is merely quaint today. This is not to say that some scientists do not still adhere to the notion that genes specify the wiring of the brain—many still do. But the past thirty years have revealed much about the development of the brain and its fine structure, and it has become clear to those who work in this area that regardless of how one defines information, there is not enough of it in the genes to wire a brain. Of course, we should not be surprised by this given that genes cannot even specify how to fold a protein.

If genes actually specified the structure of the brain and the wiring among its neurons, one might expect that brain development would look something like the construction of a house, in which the contractor orders only those materials that are necessary to accommodate their use as prescribed in the blueprint. In contrast, human brain development is characterized by an enormous proliferation of neurons—as many as 250,000 per minute—a large proportion of which will subsequently die as they compete among themselves for connections with other neurons and with muscle (as with the bulbocavernosus system described earlier); neurons that form strong connections do not die. In addition, all neurons establish an excess of contacts with other neurons and with muscle fibers that are eventually pruned back to form the more precise functional relations that characterize the adult nervous system. Thus, as brain development proceeds, the web of neuronal connections becomes increasingly focused as neurons die and as neural connections converge on their targets.

Eventually, the brain represents our various sensory systems in the form of maps. For example, the skin surface of the human body is mapped onto the surface of the cerebral cortex such that adjacent regions of skin are sensed by adjacent regions of cortex. Maintenance of these neural maps requires constant updating and calibrating, even in adults. Brain development, then, is not only dynamic, it is highly competitive and, just as Adam Smith envisioned for the economy, it is this competition within each of our brains that produces the fine-tuning of nervous system function. In the words of neurobiologist Dale Purves, "neural activity modulates the growth of nerve cells and neural circuits, ultimately affecting the overall growth of the brain and its constituent parts. By modulating neuronal growth, the activity associated with experience permanently alters the circuitry of the developing brain, thereby storing information."

What, then, is the role of genes in brain development? As with every other part of the body, genes are a critical part of the causal cascade of brain development. Today, we know that gene mutations can result in highly specific changes in neural architecture, thereby providing researchers with valuable tools to probe brain structure and function. Viewed in this light, manipulating genes is a useful tool for investigating the nervous system, but we should not carelessly leap from such observations to the unwarranted conclusion that these mutations provide insight into the genetic program for development—such a program simply does not exist. Returning to Gunther Stent, writing in 1981 at a time when the explosion of interest in activity-dependent processes in neural development was just beginning, we find a particularly concise statement of the role of genes in brain development:

> . . . the viewpoint that the structure and function of the nervous system of an animal is specified by its genes provides too narrow a context for actually understanding developmental processes and thus sets a goal for the genetic approach that is unlikely to be reached. Here 'too narrow' is not to mean that a belief in genetic specification of the nervous system necessarily implies a lack of awareness that in development there occurs an interaction between gene and environment. Rather, 'too narrow' means that the role of the genes, which, thanks to the achievements of molecular biology, we now know to be the specification of the primary structure of protein molecules, is at *too many removes* from the processes that actually build nerve cells and specify neural circuits which underlie behavior to provide an appropriate conceptual framework for posing the developmental questions that need to be answered.

In other words, problems arise in the study of development when we attempt to explain a phenomenon that is beyond our reach—*too many removes* from the problem at hand. This overreaching can occur across time or space but, regardless, we can protect ourselves from this danger by always seeking local, mechanistic connections between what we wish to understand and the methods we use to understand it. As we will now see, maintaining a focus on local interactions has provided surprising insights into the development of the visual system.

The processing of visual information begins with the entry of light into the eyes. Each eye is, in effect, a miniature brain that detects light within a layer of neural tissue—the retina—and performs some preliminary computations. Neural signals from the retina are then sent through a bundle of fibers—the optic nerve—to the brain, where this output is further analyzed in stages. One stage occurs within the visual cortex, a part of the brain that you can locate by patting yourself on the back of your head. It is here that complex visual scenes are broken down into their individual elements—color, movement, and location in space—before being recombined to create a seemless perception of the visual world.

In humans and many other mammals, as well as many predatory birds, the eyes are located at the front of the head, an arrangement that allows for binocular vision and thus the ability to perceive the world in three dimensions. To produce binocular vision, each eye sends information to both sides of the brain such that the brain represents the central part of visual space twice, but from the slightly different perspectives provided by each eye. This means that the brain must keep track of which visual information is coming from which eye. As David Hubel and Torsten Wiesel demonstrated in their groundbreaking experiments beginning in the 1960s, when a special dye that travels along nerves is injected into one eye and the visual cortex is later prepared to reveal the location of the dye, the visual cortex reveals a zebra-like arrangement of alternating stripes, or columns. These *ocular dominance columns* reflect the orderly, alternating arrangement of neural connections from each eye to the brain. But how do these ocular dominance columns arise?

In many mammals, the ocular dominance columns are not present at birth but only develop over the ensuing weeks or months. Thus, one might wonder whether postnatal experience with light is necessary for their development. This is not the case, however, because isolating animals from visual experience by rearing them in total darkness does not prevent the development of columns. Based on this finding, if ocular dominance columns were a behavioral rather than an anatomical feature, we might conclude that they are instinctive, innate, hardwired, and genetically determined.

Remember, however, Lehrman's assessment of the isolation experiment, central to Lorenz's approach to identifying instinct: the "important question is not, 'Is the animal isolated?' but '*From what* is the animal isolated?'" With respect to the development of ocular dominance columns, the key is to be absolutely clear as to what we mean by visual experience. Most of us would think that visual experience refers to what happens when we view the world through open eyes. As far as the brain is concerned, however, it does not matter if the retinal activity it is detecting is caused by the actual entry of light into the eye or by little elves applying electric charges to the optic nerve. Of course, there are no elves; instead, neurons within the retina are spontaneously active, which means that they become active even in total darkness. Activity in these neurons—which begins in the fetus—contributes to the development of ocular dominance columns and helps to explain how these columns begin to develop prenatally. When the spontaneous activity of retinal neurons is silenced, the development of ocular dominance columns is prevented. Thus, in this case, the experience essential to visual system development is self-produced, a classic example of how the nervous system develops by pulling itself up by its bootstraps.

Although we now know that spontaneous activity of retinal neurons can substitute for light in the development of ocular dominance columns, it is still unclear how and why the visual cortex responds to this incoming retinal activity by producing columns. Is there something special, then, about the visual cortex of mammals that makes it capable of producing columns? Is there an ocular dominance columns gene?

To answer these questions, recall that, in many mammals, each optic nerve emerging from each eye projects to both sides of the brain. In contrast, frogs and many other animals lack binocular vision because their eyes are spaced widely on opposite sides of the head, providing the two eyes with nonoverlapping views of their visual world; in these animals, the optic nerve from each eye projects entirely to the opposite side of the brain and, importantly, these animals do not have ocular dominance columns. What would happen, however, if the eyes of frogs could be moved forward so that they were positioned more like that of a mammal's? If this rearrangement were accomplished early enough in development, then we could determine whether ocular dominance columns develop as a consequence of the forward positioning of the eyes and the subsequent competition of the optic nerves from the two eyes for access to the same parts of the brain. Leaving aside the oddity of warping a frog in this way, it seems like the height of naïveté to imagine that such an exquisite feature of higher brain organization could be produced by such a crude rearrangement of the external features of the frog.

To produce competition between the nerves emerging from a frog's eyes, experimenters surgically manipulated an embryonic frog so that it developed a third eye on its forehead. The optic nerve of the retina of this third eye grew toward the brain and made connections similar to those of the other two eyes; importantly, these connections from the third eye were made on both sides of the brain, thus producing competition with the nerves arising from the other two eyes (as occurs normally in mammals). Amazingly, out of this artificially produced competition between optic nerves arose the development of alternating bands of tissue that resembled the ocular dominance columns of mammals. Thus, ocular dominance columns can be engineered in a species that normally does not produce them through a simple change in the peripheral anatomy. No genetic manipulations are required.

The engineering of a three-eyed frog was accomplished by implanting tissue from the eye region of one embryo (the donor) onto the forehead of another embryo (the host). Such manipulations are hit-and-miss, requiring the experimenter to identify the right donor tissue and the right host location at the right time during development. Windows of opportunity for these maneuvers are narrow, as the malleability of embryonic tissue lessens with time. The hits, however, have provided striking insights into induction, that is, the process by which one region or layer of embryonic tissue shapes the development of an adjacent region or layer.

Appreciating the nature of induction is straightforward when one is reminded that all cells within an embryo have the same complement of DNA but that the DNA of each cell is expressed differently, depending, for example, upon the DNA's location within the embryo. As we have seen, the cytoplasm plays a crucial inductive role at the earliest embryonic stages and inductive interactions continue throughout development between abutting embryonic tissues. In other words, from the perspective of embryonic tissue, induction encapsulates the notion of local experience shaping future development and is central to our understanding of how complex behaviors, such as instincts, develop.

As an example of induction and its centrality to understanding developmental process in all its curvy complexity, consider the following puzzle: a researcher studying the organization of the visual system notices in Siamese cats that the neural projections from the retina to each side of the brain are oddly disrupted. Whereas in most breeds of cats approximately half of the projections from each eye cross over to the other side of the brain for processing, much more than half of the projections cross over in Siamese cats. The result of this miswiring is bungled connectivity within the parts of the brain that produce vision and, in some instances, crossed eyes as well. Wondering if the problem is related in some way to the fact that Siamese cats are albinos, the researcher performs similar experiments using albino rats, white tigers, and pearl mink and finds similar abnormalities in the wiring of their visual systems. But what could possibly be the link between a genetic abnormality that deprives fur (and, in humans, skin) of melanin and the selective disruption of visual system development? Does the single absent gene serve double duty, or is the absence of this gene somehow predictive of the absence of a second gene that is the true cause of the brain maldevelopment?

When Gunther Stent states that the development of the nervous system is "a historical phenomenon under which one thing simply leads to another," he is providing a clue to how we should approach all such developmental mysteries. We could, of course, label the Siamese cat's visual problems as "genetic" and leave it at that, but we would lose so much by doing so. If, however, we work the problem historically, we are led to consider the possibility that the Siamese cat's visual problems might begin at the source of the optic nerves—the eyes. Thus, perhaps the link between albinism and visual system maldevelopment has something to do with the fact that most eyes, like fur, are pigmented, whereas the eyes of Siamese cats are unpigmented (giving them a bluish appearance). But how could eye color possibly influence the growth of the optic nerve?

There are two tissue layers in the back of the eye: the pigment epithelium—the layer at the very back that gives eyes their color—and the so-called neural retina—the layer in front of the pigment epithelium that contains the photoreceptors and other neurons that give rise to the optic nerve. Direct evidence of inductive processes in the eye comes from studies of a mutant rat that exhibits selective degeneration of photoreceptors; this mutation, however, does not affect the neural retina or photoreceptors directly, but rather causes a defect in the pigment epithelium that, in turn, disrupts the normal induction of the neural retina. Here, then, lies the explanation for how eye color relates to the development of the nervous system: loss of pigmentation in the eyes of Siamese cats disrupts the normal inductive interactions that contribute to the development of photoreceptors. Consequently, normal development of the optic nerve is disrupted, resulting in abnormal brain development and related visual system disturbances.

▪ ▪ ▪

The half-century since Lehrman's assault on Lorenz has produced extraordinary support for the epigenetic approach to the development of complex behavior. We may all agree that animals, including humans, have instincts in the sense that they exhibit complex, functional, species-typical behaviors. How we explain these behaviors, however, is at the heart of the matter.

Some speak of genetic control over development, but research on Bicoid protein in flies, phenylketonuria in humans, temperature-dependent sex determination in reptiles, intrauterine position in rodents, and methylation patterns in embryos show that the line of demarcation between genetic and environmental control fades the closer one inspects developmental process in action.

Some view instincts as emerging independent of experience, but the work of Celia Moore and others on the developmental consequences of anogenital licking shows that the contributions of experience to behavioral development are often indirect and nonobvious.

Some argue that genetic change is the sole mechanism of heredity, but the work of Michael Meaney and others shows that significant modifications in behavior can be passed from one generation to the next through non-genetic modes of inheritance.

Finally, some believe that brains are genetically hardwired to produce instinctive behaviors, but we now know that genes operate, in the words of Gunther Stent, at "too many removes from the processes that actually build nerve cells and specify neural circuits which underlie behavior."

The lesson is clear. To move beyond a superficial understanding of instinct, we should never be satisfied with explanations that gloss over the explanatory chasm that separates genes and behavior. As we will see in the next chapter, spanning this chasm requires local, mechanistic analyses at each stage of development.

One thing leads to another.

# 5

# DEVELOPING AN INSTINCT

IN 1802, ERASMUS DARWIN DIED, seven years before his grandson Charles was born. Like Charles, Erasmus was a famed author and scientific thinker, as well as an evolutionist who struggled to understand animal behavior without appealing to divine intervention. Whereas natural theologians such as William Paley, Erasmus's contemporary, defined instincts as divinely inspired behaviors that are expressed before an animal has gained any relevant prior experience, the elder Darwin argued that neither animals nor humans exhibit blind instinct; rather, he argued, experience and reason guide animal behavior, just as they guide human behavior. So committed was Erasmus Darwin to the notion that instincts, as commonly construed, do not exist, that he was led to suggest that rational behaviors expressed at birth or hatching must be shaped prenatally by experiences within the fetal environment. For example, he attributed the coordinated pecking of newly hatched chicks to the mouthing experience gained while still in the egg.

William Paley's and Erasmus Darwin's perspectives on behavior and cognition are complex amalgams of various positions. For example, Paley was committed to the concept of instinct, to clear lines of demarcation between man and brute, and to a firm belief in spiritual guidance, whereas Erasmus Darwin rejected the concept of instinct entirely, believed in mental continuity between humans and other animals, and considered mental processes to be firmly grounded in the material world. Subsequent thinkers have mixed and matched these opinions to suit their own beliefs. Indeed, Erasmus Darwin's own grandson was committed to the notion of mental continuity but did not reject the concept of instinct, and most nativists today espouse a belief in instinct without any hint of religious fervor.

No, God is not the issue for nativists, but rational, intelligent design is. For example, two prominent proponents of nativism within the field of developmental psychology, Elizabeth Spelke and Elissa Newport, have recently written of their desire to convince us "that questions about what is innate and what is learned are as meaningful as our ancestors thought they were." The ancestors to whom they refer are the great rationalist philosophers, beginning with Plato and continuing through Descartes and Kant, who relied on reason and logic alone to address the big questions concerning the origins of knowledge.

One developmental psychobiologist, Timothy Johnston, has noted that the rationalistic tradition addresses the question of behavioral development by looking for "precursors that are rationally related to the behaviors for which they are supposedly responsible." Followers of this tradition expect that the developmental precursors of instincts will reveal themselves through their relevance, obviousness, and transparency as developmental precursors. Johnston contrasts this tradition with the natural history tradition which is thoroughly grounded in empiricism and "aims to describe the brute facts of natural phenomena and then tries to explain them." Thus, while rationalists expect development to follow a course that a rational, designing mind would chart, empiricists have no such expectations. In other words, although the rationalist tradition glorifies the power of the mind, it blinds us to the nonobvious, meandering, and nonrational paths that typify behavioral development.

▪  ▪  ▪

You may recall that Charles Darwin, in *The Origin of Species,* defined instinctive behavior as behavior that is performed by many animals in the same way, without experience, and without knowing its purpose. Building on Darwin's interest in instincts as products of evolution, Konrad Lorenz wrote that instincts "form a natural unit of heredity. The majority of them change but slowly with evolution in the species and stubbornly resist learning in the individual."

Both Darwin and Lorenz placed a heavy burden on this single word: instinct, meaning inborn, experience-independent, stereotypic, species-typical, inherited, learning-resistant, and unmodified by ongoing stimulation. But why, we might ask, would nature prefer instincts to have all of these characteristics? Why must a species-typical behavior also be inborn and resistant to learning or context? What principle is being served and what cause is being promoted by such rigid, dichotomous conceptions of behavior?

Lorenz's cause was the selling of a new conceptualization of behavior, one that would place the study of behavior on firm scientific footing. In contrast with the psychologists and their "preoccupation with external influences on behavior," Lorenz sought the "inner structure of inherited behavior," the ancient kernels of behavior that reveal the shared genetic heritage of widely divergent species. One notorious example favored by Lorenz concerns the way that dogs and birds scratch themselves. As most of us have observed, dogs scratch themselves by steadying themselves on two forelegs and one hindleg, while the other hindleg reaches over a foreleg to accomplish the scratch. Writing in 1958, Lorenz conveys his wonder that

> most birds (as well as virtually all mammals and reptiles) scratch with precisely the same motion! A bird also scratches with a hindlimb (that is, its claw), and in doing so it lowers its wing and

reaches its claw forward in front of its shoulder. One might think that it would be simpler for the bird to move its claw directly to its head without moving its wing, which lies folded out of the way on its back. *I do not see how to explain this clumsy action unless we admit that it is inborn.* Before the bird can scratch, it must reconstruct the old spatial relationship of the limbs of the four-legged common ancestor which it shares with mammals.

Although head-scratching seems like a trivial behavior, the apparent inefficiency of overwing head-scratching in birds provided powerful support for Lorenz's core contention that the behavior of modern animals is determined by ancient inheritances. Like those who are mesmerized by the spooky similarities of identical twins reared apart, Lorenz was mesmerized by the superficially similar head-scratching behavior of dogs and birds and could not imagine any alternative but that the behavior is inborn in both and is "part of their genetic heritage." We should not, however, confuse a failure of imagination with sound argument.

In fact, research in numerous avian species has shown that, contrary to Lorenz's strongly held belief, the underwing method of head-scratching is evolutionarily more ancient than the overwing method. Moreover, some birds, such as the wood warbler, begin life using the underwing method and only switch to the overwing method as they get older. Perhaps even more surprising is the finding that adult swallows use the overwing method while perched, but switch to the underwing method while in flight, a strong indication that this behavior is surprisingly context-dependent. On the basis of all the accumulated evidence, the zoologists Edward Burtt and Jack Hailman suggested that birds may exhibit the overwing head-scratch for reasons having to do with posture, balance, and their center of gravity; when these conditions change, so too may the method of head-scratching.

Lorenz was driven to his extreme position in part by a desire to distance himself from psychologists and their seeming fixation on learning. As a consequence, he focused on the antithesis of learning, namely, those behaviors that are inborn, genetically determined, rigid, and unlearned. As we have just seen, however, even head-scratching does not fit easily into these dichotomous categories.

Fifty years ago, some recognized that it was time to end such fruitless black-and-white debates between what is inborn and what is learned. Frank Beach, one of the founders of behavioral endocrinology (the field that examines the relations between hormones and behavior), sought a compromise between the two extreme positions while calling attention to the paucity of information on the ontogeny—the development—of behavior. "No bit of behavior," he wrote in 1955, "can ever be fully understood until its ontogenesis has been described." Beach believed that with such developmental analyses of behavior, coupled with thoughtful examination of the relationships between genes and behavior, "the concept of instinct will disappear." Obviously, a half-century later, the concept of instinct has not disappeared; nonetheless, as we will see in the remainder of this chapter, advances in our understanding of behavioral development are reshaping the meaning of instinct.

### *How an Instinct Is Learned*

The arguments of epigeneticists like Lehrman, Beach, and others had an impact. One young ethologist, Jack Hailman, embarked on a program of research in the early 1960s that addressed the developmental and experiential origins of instinctive behavior. In a paper provocatively entitled "How an Instinct Is Learned," Hailman noted some interesting parallels between instincts and other behaviors that are typically not considered instinctive. For example, he observed that "braking an automobile and swinging a baseball bat are complex, stereotyped behavioral patterns that can be observed in many members of the human species, and these patterns certainly cannot be acquired without experience." So, Hailman wondered, "perhaps instincts are at least partly learned."

Although Hailman's research questions were new to ethology, the subject of his research—pecking gulls—was not. Pecking is a feeding behavior, a basic element of the gull's behavioral repertoire that is necessary to meet its survival needs. It also appears to possess all of the basic features of an instinct, including its independence from learning. How else to explain the following observation of an adult laughing gull interacting with its offspring?

> The parent lowers its head and points its beak downward in front of a week-old chick. If some time has passed since the last feeding, the chick will aim a complexly coordinated pecking motion at the bill of the parent, grasping the bill and stroking it downward. After repeated pecking the parent regurgitates partly digested food. The pecking motion of the chick is thus seen to be a form of begging for food. If one watches further, one sees the chick peck at the food, tearing pieces away and swallowing them. Pecking is therefore also a feeding action.

Thus, pecking serves a dual purpose: when aimed at the parent's beak, it elicits regurgitation of food and when aimed at the food itself, it leads to ingestion. But where does the chick's "complexly coordinated" pecking behavior come from?

The finely honed feeding behavior of a chick, only days old, would lead many to adopt the straightforward notion that chicks behave as they do because natural selection provided them with instincts—that is, complex motor and perceptual skills that promote survival immediately upon hatching. Looked at in this way, instinct provides a way to bypass the need for the uncertain and time-consuming learning process, a way to supply the chick with the answers to the exam before it is administered. Simply put, chicks don't need no education.

Or do they?

Hailman began his consideration of pecking by testing chicks as close to the time of hatching as possible. Perhaps, he thought, even a few days of post-hatching experience were enough to significantly enhance the chick's behavior. Turning first to the accuracy of pecking, he painted the image of the head of a parent onto small cards and allowed the chicks to take target practice at them. The chicks' performance on the day of hatching and on several days thereafter was scored, revealing a striking increase in performance accuracy. Specifically, while chicks successfully pecked at the target beak with an accuracy of 33 percent on the day of hatching, their accuracy increased to 75 percent only two days later. Thus, only two days of experience were sufficient for chicks to become sharpshooters. But what kind of experience was needed for this improvement in accuracy? Did chicks improve their accuracy by honing their beak-eye coordination, or was any kind of experience outside the egg sufficient?

Hailman addressed these questions by manipulating the early experience of chicks. For example, he raised some chicks in total darkness and assessed their pecking accuracy a few days later. Interestingly, chicks that were denied early visual experience showed a significant stunting of pecking accuracy. Nonetheless, even dark-reared chicks improved their pecking performance somewhat over the first few days, suggesting that some improvement was possible without visual experience. One might be tempted to conclude from this that experience is unnecessary for improvements in pecking performance, perhaps reflecting an instinctive "pecking enhancement program." But, once again, recall Lehrman's critical question—"*From what* is the animal isolated?"—when considering the vaunted isolation experiment. Yes, Hailman showed that visual experience is not essential for some improvement to take place. But there may still be other experiential factors that may have contributed to the chick's improved performance, such as improvements in its ability to steady itself on two spindly legs. After all, a steady aim requires a steady stance.

Hailman performed many other experiments to examine the development of feeding behavior in chicks. For example, he examined which features of the parent reliably elicit pecking responses. Using a variety of models of laughing gull and non-laughing gull heads and beaks, Hailman found that chicks were most sensitive to the features of the parental beak (and not the whole head). But much to his surprise, laughing gull chicks pecked just as readily at the model of a herring gull as at a model of its own species, despite the fact that these two species do not resemble each other. A week of experience with its own parents, however, was enough for a chick to develop a preference for models of its own species, as well as a keen sensitivity to even small modifications in the details of head and beak.

Clearly, newly hatched chicks focus predominantly on beaks and not heads when targeting their pecks. But what is it about beaks that is so attractive? It appeared that neither the beak's color nor its shape matters much. Perhaps, Hailman wondered, the movements of the beak in front of the chick are important. He tested this hypothesis by moving red wooden dowels in front of chicks and monitoring their pecks and he found that chicks preferred dowels with a width of eight millimeters that moved horizontally at a speed of twelve centimeters per second. Remarkably, these movements match the chick's-eye view of the parent's actual behavior. Clearly, because these were newly hatched chicks, post-hatching experience could not account for this match. Hailman thus concluded from these experiments that, contrary to previous assumptions about pecking behavior, "the newly hatched chick responds best to a very simple stimulus situation," a situation in which "the characteristics of the parent match the chick's ideal more closely than any other object in the environment."

But what does Hailman mean by "the chick's ideal?" Where does this ideal, simple as it may be, come from? Have we finally arrived at the core instinct that guides pecking in the chick?

Before we get carried away, we should remember where we were at the beginning of this discussion. Faced with the "complexly coordinated" pecking behavior of a young chick, the temptation to cast the umbrella of instinct over the entire behavioral repertoire was almost irresistible. But then we saw how a critical mind and precise experimentation could accurately reveal the developmental process that gives rise to complex, adaptive behavior. It was in this light that Hailman chose to frame his findings, emphasizing how the "chick begins life with a clumsily coordinated, poorly aimed peck motivated by hunger and elicited by simple stimulus properties of shape and movement." Clearly, this behavior does not satisfy the traditional ethological definition of instinct: that it is functional when first expressed, that experience is not a necessary prerequisite for its expression, and, indeed, that it stubbornly resists learning.

Thus, Hailman's message is that species-typical, adaptive behaviors develop in complex ways that belie the rigid definitions that were bandied about with such confidence. In the case of pecking, a chick requires experience to steady its stance, perfect its aim, and recognize its parent. That chicks behave on a steep learning curve is not irrelevant here, and one may even wish to argue that the steepness of the learning curve itself is an instinct. But to make such an argument is to admit— against the ethological tradition of Lorenz and in line with Hailman, Schneirla, and Lehrman—that instinct, learning, and other forms of experience are all intricately interwoven. And that is exactly the point.

Hailman concluded his article with a single paragraph that succinctly integrates many of the issues that we have examined thus far. He writes that his examination of pecking behavior in gulls

> strongly suggests that the normal development of other instincts entails a component of learning. *It is necessary only that the learning process be highly alike in all members of the species for a stereotyped, species-common behavioral pattern to emerge.* The example of the gulls also shows clearly that behavior cannot meaningfully be separated into unlearned and learned components, nor can a certain percentage of behavior be

attributed to learning. Behavioral development is a mosaic created by continuing interaction of the developing organism and its environment.

This statement beautifully expresses the sufficiency of common early experiences as formative influences in the production of species-typical behaviors, the pitfalls of false dichotomies, and the dynamism of the developmental process.

But if the first pecks of a chick do not qualify as an instinct, then what should we call them? Many years before Hailman, the pioneering Russian physiologist Ivan Pavlov used pecking to illustrate his view that many of the so-called instincts would be better referred to as reflexes:

This little creature [that is, the chick] reacts by pecking to any stimulus that catches the eye, whether it be a real object or only a stain in the surface it is walking upon. In what way shall we say this differs from the inclining of the head, the closing of the lids, when something flicks past its eyes? We should call this last a defensive reflex, but the first has been termed a feeding instinct: although in pecking nothing but an inclination of the head and movement of the beak occurs.

Pavlov continued with this line of reasoning by undermining the notion that instincts are an obviously higher form of behavior than reflexes. It is said, Pavlov noted, that instincts are more complex than reflexes, but vomiting requires the complex coordination of many muscle groups and yet is called a reflex. Similarly, Pavlov pointed out that, as is commonly assumed for instincts, reflexes are also often linked together to form complex chains, are modified by the internal state of the organism, and involve coordinated activity throughout the whole animal. Faced with two terms between which "we are puzzled to find any line of demarcation," Pavlov expressed a preference that *reflex* should be the preferred term because "it has been used from the very beginning with a strictly scientific connotation."

Still, there is something unsatisfying about all of this. Words and arguments are one thing, but a newly hatched chick aiming a peck, even a clumsy one, at its parent's beak is still a marvel. Call it an instinct. Call it a reflex. Call it whatever you want. We still—dare I say instinctively—fixate upon it and demand an explanation.

What form will that explanation take? Although Hailman pushed his research back to the moment of hatching, behavior begins in the embryo. It is in the embryo, then, that we must look for the developmental origins of behavior.

### *Assembly Required*

Animals enter the world exhibiting varying degrees of maturity that reflect different levels of parental care. At one end of the spectrum are those mammals and birds born in a very mature, or precocial, state—horses, elephants, guinea pigs, ducks, and quail—with open eyes and the capacity to move about their world with little or no assistance from a parent. At the other end of the spectrum are species in which newborns are born in a very immature, or altricial, state. Having worked with infant rats for many years, I am all-too-familiar with their small, furless, pink bodies, sealed eyelids and ear canals, and limited motor skills. Puppies and kittens also fall into this category, as do those species of birds, such as owls and canaries, that begin their lives incapable of flight and restricted to their nest, awaiting the next portion of regurgitated food to be deposited into their upturned, gaping mouths.

The division between precociality and altriciality is fuzzy because there are many dimensions on which to judge the maturity of a young animal. Humans exemplify this fuzziness in that they are not easily assigned to either extreme, exhibiting well-developed sensory capabilities at birth coupled with limited motor skills. If a mother turns her head for a moment, it is unlikely that her newborn will have traveled very far.

Ducklings, however, are another story. As highly precocial young, ducklings emerge from the egg ready to roll. Although extremely mobile, they still require parental care and protection. It is not surprising, then, that ducks, and other species of precocial birds, such as geese and chickens, have evolved mechanisms that promote the maintenance of close proximity between parents and their young. Specifically, imprinting mechanisms ensure that a duckling, upon emergence from its egg, will follow the first moving object that it sees. In the natural environment, this is typically the mother.

Whereas it was Oskar Heinroth who first demonstrated in 1910 that a duckling would readily imprint on a human, it was Konrad Lorenz, beginning in the 1930s, who used imprinting and other instinctive phenomena as the empirical foundation for the new science of ethology. For Lorenz, imprinting exhibited a number of unique features that distinguished it from more conventional learned behaviors. For example, he noted that there exists a brief "critical period" soon after hatching during which the window for imprinting remains open (the preferred term today is *sensitive period* to denote that the window for learning is much wider than first thought). Once imprinting occurred, Lorenz believed, an irreversible preference was established that could not be superseded by subsequent experience. Furthermore, this early imprint was thought to last into adulthood as a guide to ensure the selection of an appropriate mate (a phenomenon called sexual imprinting).

The various modes of learning and conditioning then being promulgated by psychologists, Lorenz argued, were not able to account for the rapidity, specificity, and durability of the imprinting process. As ethology developed, a variety of metaphors emerged to better communicate the uniqueness of imprinting and related behaviors (such as song learning in birds). For example, it was suggested that birds possess a learning program, or that they arrive in the world with a mental tape recorder waiting to be turned on, or slots waiting to be filled, or internal templates to be matched with appropriate stimuli in the outside world. But do these metaphors accurately capture the reality of imprinting?

Careful work by many investigators has revealed that what Lorenz conceptualized as a unitary and genetically determined behavior comprises, in fact, two processes, each of which is dependent upon experience. First, as we have already seen, newly hatched chicks exhibit a predisposition to approach objects that resemble the mother. Second, the actual filial imprinting process entails a developing preference for an object that is shaped by prior exposure to that object. Remarkably, although these two processes result in filial attachment under natural conditions, laboratory work has shown that they are not inextricably linked. Indeed, these two processes appear to rely upon different neural systems located in different regions of the brain.

To study these phenomena in the laboratory, chicks are presented with natural and artificial stimuli while they walk on a rotating wheel. Using this apparatus, investigators can train a chick using one stimulus (for example, a rotating red box) and, in subsequent testing trials, they can determine whether the chick prefers the training stimulus or a brand new stimulus (for example, a rotating blue cylinder). When tested in this way, chicks typically exhibit a preference for the stimulus with which they have been trained. This apparatus can also be used to measure the strength of a predisposition for a particular visual stimulus. Thus, jungle fowl chicks will walk more when shown a stuffed jungle fowl than when shown a rotating red box.

The nature of the predisposition has been explored by measuring the preferences of chicks for various visual features of a stuffed hen. By cutting and scrambling the stuffed object and testing the chicks' preferences, it was shown that chicks respond as readily to the head as they do to the entire hen (a finding that is reminiscent of Hailman's finding with pecking in gulls). Further work showed that chicks were responding to some general geometric feature of the head because they exhibited equal preferences for the heads of other species. But lest one assume that head-like objects are uniquely preferred by chicks, it is important to point out that chicks exhibit preferences for numerous arbitrary colors, shapes, and light patterns for which an adaptive story cannot be weaved. In other words, chicks exhibit a variety of predispositions to visual stimuli, only some of which seem to be connected with the adaptive outcome of filial imprinting.

But where do predispositions come from? It is known, for example, that a chick reared in total darkness still develops a predisposition to approach a stuffed hen, thereby ruling out a role for visual experience. On the other hand, chicks reared in total darkness will develop the predisposition only if they receive one of a number of non-specific experiences, such as running in a wheel or exposure to the hen's maternal assembly call. In other words, non-specific, *non-visual* factors can promote development of a *visual* predisposition, even though there is no obvious or rational relationship between the predisposition and the form of the experience.

As mentioned earlier, filial imprinting entails more than merely a predisposition to approach an object that resembles the mother; it also entails a preference for that object that, under normal circumstances, builds on the predisposition but is not dependent upon it. To appreciate this independence, one need only consider that chicks tend to approach objects with which they are familiar, regardless of whether that familiar object is a stuffed hen or a rotating red box. Chicks also exhibit an enhanced preference for a visual form (such as a red box) when it has been previously associated with the hen's maternal call, consistent with well-known principles of associative learning. Even more surprising, predispositions for one object do not preclude a learned preference for a second object, and such learned preferences do not overshadow the original predisposition. Clearly, filial imprinting is exceedingly more complex than originally imagined by the early ethologists, including Lorenz.

Upon consideration of the many studies of filial imprinting in young birds, the ethologist Carel ten Cate concludes that filial imprinting results from a complex combination of pre-hatching and post-hatching experience:

> This complexity will not exactly cover what most people have in mind when they mention filial imprinting as an example of 'template' or 'programmed' learning, nor is it covered by 'slots' waiting to be filled. In contrast, the apparent programmed nature of the developmental process originates from an intricate interplay between organism and environment, and the outcome is dependent on environmental stimulation as much as on changing properties of the organism itself.

In addition, the totality of research on filial imprinting calls into question, once again, the wisdom of drawing stark distinctions between learning and other forms of developmental experience.

As fascinating and complex as visual imprinting may be, it is but one facet of the process by which hens assemble their young and maintain group cohesion. As discussed above, hens produce maternal assembly calls that can facilitate the emergence of a visual preference. When the behavioral embryologist Gilbert Gottlieb began studying the responses of chicks and ducklings to these calls in the 1960s, he made an interesting discovery. As he reminded his colleagues at a conference in England in 1971, he had found that "hatchlings were capable of identifying the maternal assembly call of their respective species in advance of contact with the hen. Thus, learning the visual characteristics of their species through contact with the hen [that is, visual imprinting, as discussed above] was a subsidiary process that took place in the context of the hatchlings' attraction to the maternal call." Of course, Gottlieb's results also provided evidence that hatchlings have an inborn preference for the maternal call of its own species. This is, naturally, what most researchers concluded at the time.

But Gottlieb's intellectual influences were rather unconventional. Due in part to his interactions with Schneirla and with the Chinese behavioral embryologist Zing-Yang Kuo, Gottlieb was not a fan of conventional instinct theory and its commitment to genetically predetermined behavior. Instead, he had embraced the epigenetic idea that the behavioral capabilities of an animal, including instincts, emerge over the course of development. According to his perspective, while natural selection helps us to understand how a complex, adaptive behavior might arise in a population of animals living in a particular environment, it does not provide an answer to the developmental question of how such behaviors actually reemerge generation after generation within individual animals.

Given Gottlieb's antipathy to instinct theory and his own belief that epigenesis provided a surer path to understanding complex behavior, what happened to him next at that 1971 conference must have been horrifying. Konrad Lorenz, still a formidable man who would receive the Nobel Prize two years later, was in attendance and had just heard Gottlieb dismiss visual imprinting as a subsidiary process and seemingly argue that hatchlings have an inborn preference for their mother's assembly call. As Gottlieb recounts what happened next, Lorenz

> blithely reminded all concerned that, during development, there were two sources of information, one innate or instinctive flowing from the genes, the other flowing from the environment (imprinting in this case), and that I [that is, Gottlieb] had demonstrated the importance of the former. Lorenz had not only deftly deflected my challenge to the significance of imprinting, he had used the opportunity to say how my findings supported the nature-nurture dichotomy as well! There was some well-deserved hilarity at my expense.

Although Gottlieb did not mention it to his colleagues at the time, he had begun a series of groundbreaking experiments that would eventually give him the last laugh.

Gottlieb's discovery of an inborn preference of ducklings for the maternal assembly call surprised him. The experiment itself was simple: noting that embryos are able to hear the maternal assembly calls even while still in the egg, he incubated eggs in isolation from the mother. Having prevented his mallard ducklings and wood ducklings from hearing the maternal calls, he expected that they would fail to develop an auditory preference for their own species' call. On the contrary, he found that deprived ducklings consistently chose their own-species maternal calls over those of other species (and found the same results in jungle fowl and domestic chicks). Thus, in contrast with visual imprinting and its reliance on post-hatching visual experience, the expression of an acoustic preference for a species-typical vocalization was clearly independent of prior experience with the maternal vocalization.

Any reasonable person, and especially Lorenz, would have concluded from Gottlieb's experiment that his ducklings were exhibiting a full-blown, predetermined, genetically programmed instinct. Indeed, similar claims are made today concerning innate knowledge and behavior—in human and non-human animals—based on much less compelling evidence. Had Gottlieb been so inclined, he could have reasonably moved on to another research question, content that he had added yet another instinct story to the literature.

Once again, however, we return to Lehrman's incisive critique of the isolation experiment: The "important question is not, 'Is the animal isolated?' but '*From what* is the animal isolated?'" Addressing this question, Gottlieb noted that the incubation of eggs in isolation does not prevent all auditory experience because, beginning three days before hatching as the bill enters the air space within the egg, embryonic ducklings emit their own vocalizations! Although these embryonic vocalizations do not sound like the maternal assembly call, Gottlieb wondered whether prenatal exposure to these vocalizations could somehow induce a postnatal auditory preference. Accordingly, he showed that additional exposure of ducklings to the embryonic vocalizations enhanced their post-hatching preference for the maternal call. Demonstrating enhancement, however, does not resolve the issue of how the auditory preference is established in the first place. To resolve that issue, it was necessary to develop a procedure by which embryos could be devocalized within the egg and incubated in complete isolation from all other sounds, including the embryonic vocalizations of siblings. Once this procedure was in place, the critical experiments could begin.

Using his new procedure, Gottlieb showed that mallard embryos that had been devocalized were unable as ducklings to distinguish between mallard and chicken maternal assembly calls, even when tested as many as forty-eight hours after hatching. (These preference experiments are performed by placing chicks in a small arena, playing various maternal calls through loudspeakers, and measuring the amount of time that the chicks spend near each loudspeaker.) In contrast, ducklings that had been devocalized after they had been briefly exposed exclusively to their own vocalizations exhibited the normal, strong preference for mallard maternal calls. Thus, auditory self-stimulation was sufficient to establish an auditory preference, a truly stunning result when considered in the context of conventional instinct theory. But this was only the beginning.

One issue that lingered after Gottlieb's initial devocalization experiments was whether there were specific properties of the mallard embryonic vocalization that were necessary to establish the post-hatching preference for the distinctive features of the mallard maternal call. To address this issue, Gottlieb focused on the fact that the individual notes in each maternal call have characteristic frequency components and are emitted at characteristic repetition rates. Intriguingly, the mallard embryonic call shares two important features with the mallard maternal call: high frequency components and a relatively fast repetition rate. Thus, Gottlieb wondered whether devocalizing the mallard embryos desensitized ducklings to one or both of the characteristic features of the maternal assembly call and thereby resulted in the inability to distinguish between the mallard call and, for example, the chicken call. Furthermore, he wondered whether mere exposure to the characteristic frequencies and repetition rates of the embryonic call, as occurs under normal rearing conditions, is sufficient to establish a preference for the maternal assembly call.

To make a long and very complicated story short, Gottlieb performed numerous experiments in which embryonic calls were filtered to alter their frequencies and manipulated to alter their repetition rates. Ultimately, he demonstrated that normal development depends upon very specific experience with the distinguishing characteristics of the embryonic call. Moreover, this experience must occur during the later stages of embryonic development if the species-typical auditory preference of newly hatched ducklings is to be expressed.

How does exposure to the embryonic vocalization exert its effects on the developing nervous system to produce a post-hatching preference? Gottlieb showed that at least one mechanism entails a generalized change in auditory function; specifically, around the time of hatching, exposure to self-produced and sibling-produced embryonic calls substantially enhances the sensitivity of the auditory system. When embryos are deprived of this normal auditory experience, ducklings do not exhibit the typical increase in auditory sensitivity. Rather, as noted by Gottlieb, these ducklings "show a virtual arrest of perceptual development in the absence of normal auditory experience."

Finally, one might reasonably wonder about the malleability of the duckling's developing preference for a maternal contact call. For example, is it possible to create a duckling that exhibits a preference for the call of another species? The answer, perhaps unsurprising by now, is yes. Gottlieb demonstrated this using mallard duck embryos that had been devocalized, isolated in a silent acoustic chamber, and exposed to a tape recording of the maternal calls of other species, such as a chicken or a wood duck. After hatching, these ducklings preferred the call of the species to which they had been exposed. He even showed that mallards could be manipulated to prefer a chicken call over the maternal call of its own species.

Gottlieb's unique and detailed insights into behavioral development have placed him in an ideal position to conceptualize the many ways that experience—broadly defined—recreates complex behavior anew with each generation. Some experiences are inductive in a way similar to that discussed in the previous chapter with respect to developmental interactions at the cellular level. Induction establishes a new developmental outcome, causing development to proceed in a different direction. For example, in the absence of exposure to the embryonic call, the auditory preference for the maternal call cannot be established. Experience can also facilitate the development of a behavioral outcome and maintain a behavioral outcome once it has developed. Induction, facilitation, and maintenance comprise the basic, formative modes by which experience, working in concert with gene expression, regenerates complex, adaptive behaviors across generations.

Perhaps the most important aspect of Gottlieb's approach to understanding development is that he was the first to provide support for the notion, inspired by Schneirla, Lehrman, and others, that the causal antecedents of complex behavior are likely to be found in nonobvious places. When considering the embryonic call and its impact on auditory preferences in birds, it cannot be overemphasized that *these embryonic calls do not sound like maternal assembly calls.* Yes, Gottlieb was able to identify acoustic features within the embryonic calls that drove the subsequent development of the preference, but these features were buried within a complex acoustic stimulus and had to be carefully mined and experimentally tested. Had Gottlieb adopted the dominant view of his contemporaries concerning instincts and their predetermination by genes, the true complexities of behavioral development would be even less appreciated than they are today.

The repercussions of Gottlieb's conceptual and empirical contributions for our understanding of evolution and development are considerable:

> The search for nonobvious experiential bases to instinctive behavior is in line with evolutionary considerations. Natural selection works on adaptive behavioral phenotypes; it is completely indifferent to the particular pathway taken by the phenotype during the course of development. . . . [N]atural selection involves a selection for the entire developmental manifold. . . . Thus, nonobvious experiential involvement of a patterned kind may be much more widespread than heretofore realized. Only developmental investigations of unlearned behavior can answer that question, and that has not been a popular form of experimentation, because influential scientists have looked on seemingly innate or unlearned behavior as a predetermined outcome of development, not as a probabilistic epigenetic phenomenon. . . .

Although the full extent of Gottlieb's contributions to our understanding of behavioral development is still not fully appreciated, the clarity of his thinking and the incontrovertible nature of his empirical findings instill hope that future generations of scientists will not succumb to the temptations of the simplistic and superficial explanations of behavior that retain such overwhelming popularity today.

### *Let's Get Physical*

A central theme of this chapter, as explained through the work of Hailman and Gottlieb, has been that the early experiences of an organism need only be repeatable and predictable for behaviors to be reliably expressed across generations. Like DNA, these normally occurring experiences help to ensure that behavioral traits are reliably transmitted.

The key point to remember is that natural selection acts on the entire process—the "entire developmental manifold," in Gottlieb's words—not on any individual component of that process.

As we will see, the physical environment—influenced as it is by thermal, gravitational, mechanical, nutritional, and social factors—structures normally occurring experiences. Even self-generated behaviors by fetuses play a role. Acting and interacting in complex and often nonobvious ways, all of these factors contribute to the production of species-typical behaviors by providing opportunities for species-typical developmental experiences.

Ants and other neuter (that is, sterile) insects were of particular interest to Darwin, and he devoted many pages in the instinct chapter of *The Origin of Species* describing their remarkable behaviors. In fact, Darwin was terrified that the theoretical difficulties posed by ants—specifically, the existence of a distinct caste of workers whose sterility precludes their ability to propagate their own kind—would prove "fatal to the whole theory" of natural selection. "The great difficulty," Darwin wrote, "lies in the working ants differing widely from both the males and the fertile females in structure . . . and in instinct. . . . It may well be asked how is it possible to reconcile this case with the theory of natural selection?" Darwin's brilliant solution was to think of a colony of ants as a single entity upon which natural selection can act as a whole. Accordingly, the sterile workers contribute to the fitness of the entire colony, including their fertile relatives and, by doing so, contribute indirectly to their own propagation. Looked at in another way, Gottlieb's developmental manifold can be broadened beyond individual ants to encompass the entire ant superorganism.

This perspective of workers and queens making independent contributions to colony fitness can easily foster the notion that individual castes, with their distinct anatomy and behavior, are genetically distinct subpopulations. On the contrary, in what Bert Hölldobler and E. O.

Wilson describe as an "overriding principle," the female castes—sterile workers and fertile queens—"are differentiated by physiological rather than genetic factors." What they mean by this is that all ant larvae are genetically indistinguishable, and it is during development that particular environmental stimuli guide the larvae down a developmental path toward one caste or the other. Six such stimuli have been identified, including nutrition and temperature (which modulate growth rate, a major factor in caste determination) and the presence of a mother queen (which produces an inhibiting factor and thereby regulates the colony's caste ratio). Thus, for ants, species-typical developmental experiences are provided by environmental signals that modulate colony demographics from one generation to the next.

The resiliency of bird eggs against breaking is a double-edged sword; chicks, after all, have to get out. It is obvious, then, that hatching is a critical behavior that must be expressed by an embryo before it has ever experienced the outside world. In chickens, hatching behavior begins abruptly on the twenty-first day of incubation with rapid extensions and flexions of the legs that provide the thrusts needed to rotate the embryo within the shell as the beak cracks the egg along an internal circumference. With repeated thrusting, rotating, and cracking, a loosened cap is created that, when lifted, is the door through which the chick's head and body can emerge. The hatching movements then cease as abruptly as they began, the whole process having taken about one hour.

The movements that comprise hatching are smooth and coordinated; they are also uniquely performed at the time of hatching (although the behavioral elements that comprise hatching can be traced to seemingly functionless movements that occur earlier in embryonic life). But regardless of the origins of the hatching movements themselves, what is the trigger that initiates the entire process? Is there a hatching instinct that "turns on" when chicks reach a certain age?

Clues for answering these questions were provided when chicks, even chicks many weeks old, were placed inside glass eggs to re-establish the embryonic posture that existed just before hatching. Most significant, it turned out, is the position of the neck, bent to the right or left and held firmly in place. As Anne Bekoff and her students discovered, restoring this "embryonic" neck posture in chicks evokes the synchronized movements of the hind legs that are characteristic of hatching. Once the neck is released, these particular movements cease, and if the neck is desensitized using a local anesthetic, hatching movements do not occur at all. So, it appears that the physical bending and unbending of the neck plays an important role in the initiation and termination of hatching behaviors, respectively. But still, embryos exhibit the bent-necked position for several days before hatching actually begins. Why, then, doesn't hatching begin sooner? Although the answer to this question is not yet known, Bekoff has proposed that the physical growth of the embryo within the egg may increasingly bend and cramp the neck until "a critical point is finally reached at which the signal from the neck becomes strong enough to trigger hatching."

To be successful, infant rats must develop filial preferences and learn to recognize members of their own species. If one were to observe rat pups at fifteen days of age, one would see that they already exhibit a striking preference to affiliate with each other. When offered the opportunity to affiliate with similar-sized rodents of another species, such as a gerbil, they spurn the gerbil and choose the rat. If offered access to a warm, furry tube, the pups will again prefer to huddle with the rat. Although one might reasonably assume that natural selection would have ensured that such filial preferences, which are so critical to the pup's future social interactions, would be programmed into the pup's brain, there is no need to make such an assumption when the normally occurring experiences of rat pups provide a perfectly reliable solution.

To appreciate the elegance of this solution, recall that pups are born in a state of development that makes them particularly vulnerable to the vagaries of the external environment. Because wild rats give birth to litters of approximately six young, each pup can protect itself from the cold by huddling with its littermates; huddling helps pups to withstand the cold until, with increasing body size and the growth of fur, they can fend for themselves. For the first ten days after birth, rat pups are not particularly picky about the source of heat; they will snuggle up against other rats, of course, but also pups from other species and even inanimate objects, such as a warm tube.

Between ten and fifteen days of age, rat pups begin to prefer rat odors to those of other species. This preference emerges, however, not from the sudden maturation of a "filial preference neural module," but from the normally occurring experience of associating rat odors with warmth. Developmental psychobiologist Jeffrey Alberts and his students demonstrated this by daubing mother rats with an artificial odor during the first two postnatal weeks and then, when pups were fifteen days old, assessing their odor preferences; pups that experienced the artificial odor in a context associated with warmth—regardless of the source of that warmth— came to prefer the artificial odor over "natural" rat odor. This nearly effortless reassignment of filial preference to an artificial odor demonstrates that "rats exhibit a species-typical preference because they are reared in a species-typical environment."

Although specific odor preferences emerge in older pups, the ability to detect, orient, and move toward warmth is present almost immediately after birth. Before birth, rats, humans, and all other placental mammals develop within a temperature-controlled environment that places few thermal demands on the fetus. The maintenance of a regulated and stable thermal environment is critical for normal development; speaking loosely, we might say that the entire developmental manifold has come to expect such a cozy environment. It is for a very good reason that we caution pregnant mothers to avoid hot tubs: even small, brief shifts in temperature can have disastrous consequences for fetuses.

Thus, the temperature within the womb provides the fetus with a normally occurring experience that is essential for normal development. The same can be said for gravity. Our ability to orient our bodies in response to the force of gravity is made possible by one of the body's most complex and elegant neural and behavioral systems. At the heart of this system is the vestibular apparatus, a collection of fluid-filled tubes adjacent to the inner ear that detects changes in the acceleration and orientation of the head. The vestibular system is one of the earliest sensory systems to become functional in mammals and birds; in rats, the basic system is completed by the end of gestation and even exhibits some functionality before birth. This raises a question: Is normal stimulation of the embryonic vestibular system—either as a result of gravity or the movements of the mother—necessary for its proper development?

Most sensory systems are easy to manipulate. Light, sound, touch, taste, odors, and heat can be easily provided or denied during an experiment. In contrast, it is relatively difficult to increase or decrease gravitational forces for prolonged periods of time. Thus, scientists have begun taking advantage of the opportunities offered by the space shuttle program to test animals under prolonged conditions of weightlessness. For example, in one experiment, pregnant rats were launched on the space shuttle Atlantis and maintained under weightless conditions for eleven days spanning the ninth to the twentieth days of gestation. Upon return to Earth, the pregnant mothers gave birth within one to two days, whereupon the vestibular responses of the pups were tested.

One of the simplest tests of vestibular function in pups is to release them on their back into an aquarium filled with warm water. As they sink, pups with normally functioning vestibular systems will rotate their bodies so that they land on the bottom of the tank in a prone position (these tests are very brief and pups are quickly removed from the water unharmed). When pups gestated in space were tested soon after birth, however, Jeffrey Alberts and April Ronca found that a high proportion failed to right themselves; many of the pups floated to the bottom without flipping over. By five days of age, these space pups had caught up to the control pups, suggesting rapid readaptation to the Earth's gravitational environment. Moreover, the results of this behavioral study are consistent with those that show changes in the neural architecture of fetuses gestated or reared in microgravity. Thus, based on work that has been possible for only a few years, it appears that gravity can play an inductive, facilitative, and maintenance role for normal vestibular function.

Temperature and gravity exert their effects *on* the developing embryo. Some physical influences, however, are self-generated. For example, many of us have observed the twitching movements of our dogs and cats while in rapid eye movement (or REM) sleep. These twitches result from spontaneous, that is, self-produced neural activity within the brain and spinal cord; in fact, rapid eye movements are produced by twitches of the eye muscles (and not, as some imagine, a sign that we are watching our dreams).

Even in fetuses and newborns, spontaneous movements contribute to the development of cartilage, tendons, ligaments, and bone. Bones are highly responsive to compressive forces, whether induced by gravity or movement. In fetuses, spontaneous movements generate sufficient forces on bones to trigger cascades of gene expression that result in the production of collagen and other basic structural components. As bones and joints develop prenatally, they are sculpted and remodeled under the constant influence of fetal movements; indeed, blocking these movements using a paralytic agent prevents normal development of the skeletal system.

Neonatal motor activity may also provide critical feedback to the nervous system concerning the calibration between neuronal signals and subsequent muscle activity. The need for such calibration is self-evident. Imagine maintaining stable control over the many parts of your body as each undergoes rapid and dramatic change in relative size, as muscles grow, as neurons proliferate and die, and as neural connections are modified. The challenge posed by development is similar to the challenge of learning how to throw a medicine ball—adjusting for the added weight, using more parts of the body and more muscle groups to gain leverage, altering the throwing motion—but exceedingly more complex. Whether the spontaneous movements of fetuses played any role in such complex neuromuscular phenomena, however, was doubted by many because of the prevailing view that the job of the fetus was to wait patiently and passively for growth to happen.

A new era in the study of prenatal behavior began when researchers perfected a method for observing fetuses outside the womb without disrupting their umbilical connection to the mother. Once this task was accomplished, it was soon demonstrated that rat fetuses, within a few days of the end of their twenty-two-day gestation, exhibit a surprising array of complex behavioral patterns. For example, when lemon extract is squirted into the mouth of a fetus, it swipes its paws over its face in a coordinated fashion that resembles the facial wiping behavior typically observed in adults. Although such elicited behaviors provide a unique opportunity to examine the developmental origins of infant behavior, they do not address the meaning and importance of spontaneous behavior.

In an experiment that assessed the functional significance of spontaneous fetal behaviors, developmental psychobiologist Scott Robinson tethered the hind limbs of fetal rats on the twentieth day of gestation so that spontaneous movements in one limb resulted in the passive movement of the other. By "yoking" the limbs in this way, commands from the central nervous system to move one limb were followed immediately by feedback information that both limbs had moved. The question was whether, over time, this alteration in sensory feedback would lead to an alteration in spontaneous motor activity to the limbs. Using a sophisticated motion-analysis system to examine the simultaneity and trajectory of hind limb movements, Robinson found that fetuses experiencing only thirty minutes of hind limb yoking exhibited dramatic increases in conjugate movements, that is, movements that were initiated at the same time and directed toward the same location in space (enthusiastic sports fans make conjugate movements with their arms during a wave). Moreover, these conjugate movements continued even after the tether was cut and the limbs were again free to move independently.

These and other related experiments demonstrate that the spontaneous behaviors produced by fetuses are already being integrated within a complex system of motor output coupled with sensory feedback. The result, according to Robinson and Gale Kleven, is a dramatic reorganization of our conception of life before birth:

> In few organisms is the seemingly simple distinction between activity and experience, or between organism and environment, more blurred than in the mammalian fetus. The fetus develops in a unique environment created by tissues of both embryonic and maternal origin. It is commonplace to consider the prenatal environment as static and unchanging, and the fetus as passive and unresponsive. Research on fetal behavior and neural development is replacing this still life with a dynamic portrait of an organism that develops within and in relation to a complex and changing environment, in which the fetus is an active participant.

Rational processes have a way of revealing themselves through their obviousness and transparency. But only a rigid belief in a grand design can sustain a belief in the rationality of behavioral development. In contrast, as we have seen throughout this chapter, the path to understanding behavioral development must pass through the territory of Brute Fact, a territory inhabited by the nonobvious and the opaque. Perhaps it didn't have to be this way. But that's the way it is.

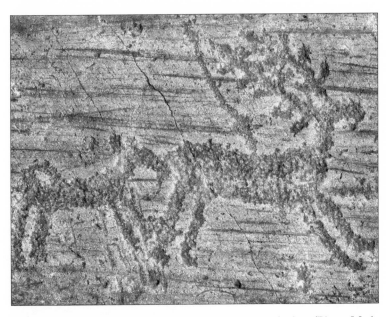

Camuni engraving depicting a dog nipping at the rump of a deer. (Photo: Mark Blumberg)

Welsh farmer with his border collie at the International Sheep Dog Trials in 1953 at Cardiff, Wales. (Courtesy of Carole Presberg and E. B. Carpenter)

A two-hour-old *Drosophila* embryo. At the anterior end of the embryo (left side in this photo) is a dark substance that forms a gradient down the embryo. This substance is Bicoid protein and its presence is necessary for the development of the fly's head. (Courtesy of Nipam H. Patel)

Danny Lehrman (Photo: Rae Silver; Courtesy of Gilbert Gottlieb)

A rat mother tending to her supine infant. (Courtesy of Jeffrey Alberts)

Gilbert Gottlieb (left), with Zing-Yang Kuo in 1963, making a window in the shell of a duck egg for observing embryonic behavior. (Photo: Gus Martin; Courtesy of Gilbert Gottlieb)

Cowbird baby being fed by an adult canary. (Photo: Andrew King;
Courtesy of Meredith West and Andrew King)

Drawing of an adult female black rat and her offspring stripping pinecones. (Artist: Bill Ferguson; Courtesy of Joseph Terkel)

A young infant showing poor judgment, plunging down a steep slope that she can't possibly traverse without falling. (Courtesy of Karen Adolph)

Katy in repose. (Photo: Mark Blumberg)

# OF HUMAN BONDAGE

THE MODERN DISCUSSION OF HUMAN instinct was given a boost over a century ago when William James devoted a chapter to the subject in his *Principles of Psychology*, widely regarded as the field's first textbook. James, trained as a medical doctor but also a gifted writer (as was his more famous brother, Henry, the novelist), conceived of instincts as the most basic of nervous system impulses; once expressed, he argued, instincts become modified by experience and lose their "blind" nature, thereafter fading in importance. Taking issue with previous thinkers on the topic, whose writings he dismissed as "ineffectual wastes of words," James conceived of instincts as the product of natural selection acting on animal, including human, behavior.

For James, the dichotomy between reason and instinct produced a false division between humans and animals. Man, he wrote, "has a far greater variety of *impulses* than any lower animal." Included among these human impulses were sucking, crying, smiling, creeping, standing, imitating, playing, resenting, emulating, sympathizing, appropriating, constructing, and hunting, as well as shyness, secretiveness, cleanliness, love, jealousy, and fear—fear of strange people, strange animals, dark places, heights, and the supernatural. Admitting that some will "find the list too large, others too small," James concluded that "no other mammal, not even the monkey, shows so large an array."

One prominent psychologist who thought James's list was too short was William McDougall, who opened and closed the first chapter of his influential *An Introduction to Social Psychology,* first published in 1908 but reprinted dozens of times throughout the century, with these words:

> The human mind has certain innate or inherited tendencies which are the essential springs or motive powers of all thought and action, whether individual or collective, and are the bases from which the character and will of individuals and of nations are gradually developed under the guidance of the intellectual faculties. . . . Take away these instinctive dispositions with their powerful impulses, and the organism would become incapable of activity of any kind; it would lie inert and motionless like a wonderful clockwork whose mainspring had been removed or a steam-engine whose fires had been drawn.

Like James, McDougall had few kind words for other writers on the topic, complaining that the words *instinct* and *instinctive* had been used "so loosely that they have almost spoilt them for scientific purposes" and, even worse, that these words provided "a cloak of ignorance when a writer attempts to explain any individual or collective action that he fails, or has not tried, to understand." With no apparent sense of irony, McDougall did nothing to break with the loose talk of the past. Indeed, what made McDougall's contribution such a crowning achievement of hubris is that, having presented himself as a man of intellectual rigor and precision, he stuffed the remainder of his book with more instincts than humans have bones. Given his penchant for free-wheeling armchair theorizing, it can be no surprise that McDougall ultimately abandoned any hope for a scientific explanation of instinct and descended into abject mysticism.

In the wake of World War I, a relatively obscure psychologist named Knight Dunlap, in an article entitled "Are There Any Instincts?," threw cold water on the instinct orgy that James and McDougall had started. Dunlap's main point in his brief article was that the grouping of behaviors into instincts "may be admitted to be a useful procedure, if it be clearly understood to be a device of convenience only, similar to the arrangement of documents in a well-ordered filing system." Taking direct aim at McDougall, Dunlap implored his colleagues to "cease talking of 'instincts.'"

Dunlap's article had little impact. In Europe, Konrad Lorenz and his fellow ethologists would soon revitalize the use of instinct. In the United States, talk of instinct did indeed fall into disfavor, but it was the rise of the school of psychology known as behaviorism, not Dunlap's article, that did the deed. Decades later, the pendulum began to swing again with the publication in 1961 of an odd little article by Keller Breland and Marian Breland. Working within the behaviorist tradition but using nontraditional animals, the Brelands were in the business of training animals to do circus-like stunts. Repeatedly, however, they seemed to hit the brick wall of instinct. For example, when they tried to train a raccoon to deposit coins into a metal box, the raccoon would not release the coins and, even more infuriating, persisted in rubbing the coins together. Attempts to train a similar behavior in a pig worked at first, but the training regimen lost its effectiveness over time as the pig began rooting the money with its snout rather than placing it in the bank.

This gradual drifting toward species-typical behaviors—behaviors that were not being explicitly reinforced—was shocking to the Brelands because, as they put it, "there was nothing in our background in behaviorism to prepare us for such gross inabilities to predict and control the behavior of animals with which we had been working for years." Stating that their animals were becoming "trapped by strong instinctive behaviors," they pronounced that when "behaviorism tossed out instinct . . . some of its power of prediction and control were lost with it." Contrary to behaviorist dogma, they argued, animals are not blank slates; species differences are important, and not all behaviors are conditioned with ease.

This was just one of many body blows that behaviorism was to absorb. Another was the insurgency of cognitivists who began talking openly about the covert mental processes that guide overt behavior. On the heels of cognitions came emotions, then feelings, then consciousness. Geneticists had something to say, as did neuroscientists, then neurogeneticists. Evolutionary biologists, anthropologists, linguists, philosophers, and computer scientists all made claims. Up sprung sociobiology, cognitive ethology, evolutionary psychology, the theory of mind movement, and a resurgence of nativism. Even the once-universally shunned anecdotal and anthropomorphic approaches of the past were being resurrected as tools for explaining animal behavior. Meanwhile, as the fruits of behaviorism continued to fuel some of the most significant advances in neuroscience, many scientists and philosophers explicitly spurned any mention of behaviorism, perhaps acting out of revenge but undoubtedly acting out of ignorance. The trends were clear: with the flood-gates opened wide, there were few forces left to constrain the wildest imaginings of modern-day McDougalls. Science is a messy business.

Unraveling the social, political, and scientific forces that have unleashed the rowdiness of the past few decades is itself the subject of another book. Thus, I will focus here on the burgeoning field of evolutionary psychology because the claims of its proponents are directly relevant to our understanding of the origins and nature of instincts.

The benefits of grounding psychology in sound evolutionary thinking cannot be doubted by anyone seriously interested in either psychology or evolution. Unfortunately, as currently practiced and preached, evolutionary psychology represents neither the best of psychological nor evolutionary thinking.

To get a sense of the general intellectual style of evolutionary psychologists, we can begin with a behavior with which we are already familiar: pecking in gulls. Recall, from the previous chapter, Jack Hailman's detailed analysis of the development of pecking behavior and compare it to this recitation by two founders of evolutionary psychology, John Tooby and Leda Cosmides. A newborn herring gull, they wrote in 1992,

> has a cognitive program that defines a red dot on the end of a beak as salient information from the environment, and that causes the chick to peck at the red dot upon perceiving it. Its mother has a cognitive program that defines pecking at her red dot as salient information from her environment, and that causes her to regurgitate food into the newborn's mouth when she perceives its pecks. These simple programs adaptively regulate how herring gulls feed their young. . . . Moreover, precise descriptions of these cognitive programs can capture the way in which information is used to generate adaptive behavior. Of course, these programs are embodied in the herring gull's neurobiological "hardware."

The stark contrast between Hailman's methodical and mechanistic insights (of which Tooby and Cosmides seem to be unaware) and the anemic summary above should be jarring.

Of course, evolutionary psychologists do not really care about gulls. They are primarily interested in humans and, more specifically, human instincts. Echoing William James and William McDougall, Tooby and Cosmides assert without evidence that "what is special about the human mind is not that it gave up 'instinct' in order to become more flexible, but that it proliferated 'instincts'—that is, content-specific problem-solving specializations—which allowed an expanding role for psychological mechanisms that are (relatively) more function-general." The next step—the one that McDougall was not in a position to take—is simply to ascribe these content-specific specializations to innately specified, hardwired, independent, neural components, or modules, each of which is conceptualized as the unbounded product of evolutionary design. But there is yet one more step to take. Not only is the human brain a "set of computational machines," but each of these machines "was designed by natural selection to solve adaptive problems faced by our hunter-gatherer ancestors." Thus, we are, like the five-thousand-year-old iceman discovered in the Alps in 1991, frozen in Neolithic time. We are bound by evolutionary descent to an ancient mode of thinking and acting. We are cavemen in designer jeans.

I have no quarrel with the basic tenets of evolutionary psychology. Of course we are the products of natural selection; of course our behavior reflects our evolutionary heritage; of course our understanding of behavior and cognition can be enriched by an evolutionary perspective. But these are hardly novel insights. What distinguishes evolutionary psychology as a movement is its commitment to a particular mode of explanation—to the belief that we are driven by our evolutionary heritage to exhibit content-specific solutions to ancestral problems. For example, evolutionary psychologists are convinced that we are innately programmed to fly into a jealous rage when jilted by a lover, to murder our step-children to relieve the burden of raising another person's progeny, or, when choosing a mate, to fixate on specific body shapes that are supposedly linked to enhanced reproductive potential. Moreover, according to Tooby and Cosmides, these innate proclivities are guided by specific neural modules:

> By adding together a face recognition module, a spatial relations module, a rigid object mechanics module, a tool-use module, a fear module, a social-exchange module, an emotion-perception module, a kin-oriented motivation module, an effort allocation and recalibration module, a child-care module, a social-inference

module, a sexual-attraction module, a semantic-inference module, a friendship module, a grammar acquisition module, a communication-pragmatics module, a theory of mind module, and so on, an architecture gains a breadth of competences that allows it to solve a wider and wider array of problems, coming to resemble, more and more, a human mind.

Used in this way, modules have little real scientific value. Their more immediate function is sociopolitical—to provide a physical face to nativism while hiding behind a patina of scientific respectability through the use of pseudo-neuroscientific ideas. But Knight Dunlap's critique of McDougall over eighty years ago is no less valid today. Filing systems are not explanatory devices.

It is important to stress that the nervous system does exhibit modularity. Each sensory system—visual, auditory, olfactory, and tactile—begins with specific receptors in the eyes, ears, nose, and skin that connect in a dedicated fashion with discrete neural systems, including distinct areas of the cerebral cortex; for example, in the extreme case of the platypus, which relies so heavily on sensory information provided by sensors in its oversized bill, two-thirds of the cerebral cortex is dedicated to processing sensory information from the bill alone. There are also discrete memory systems whose capacities and neuroanatomy reflect the adaptive specializations of different species, such as the black-capped chickadee that can remember the exact hidden locations of tens of thousands of seeds. But such legitimate invocations of modularity rest on certain knowledge regarding the neuroanatomical organization of sensory systems and discernible relationships between sensory processing and observable behaviors. In contrast, how does one specify the anatomical foundations of the semantic-inference module other than to posit that it exists somewhere in the cerebral cortex? Moreover, although part of the allure of evolutionary psychology revolves around its commitment to innateness, "there is no logically necessary connection between innateness and modularity," as Sara Shettleworth has pointed out. She continues, "The degree to which a particular aspect of information processing is influenced by experience (i.e., how it develops), whether or not it is computationally distinctive, what information it deals with, and what its current or past adaptive value might be are all separable questions." Thus, to put it bluntly, evolutionary psychologists have hijacked a legitimate concept and overloaded it with unnecessary baggage.

There is a richness to behavioral development that is not captured by evolutionary psychologists' blithe pronouncements regarding innate cognitive programs and content-specific neural modules. In the previous chapter, I focused on relatively simple behaviors and their developmental origins. I will now shift to stories of more complex behavioral processes in birds and mammals, including adult behavior. With each story, it should become increasingly clear that the various claims of evolutionary psychologists regarding the emergence of complex behavior are naïve when applied to rats, cowbirds, and moose, let alone humans.

## *You Can Lead a Horse to Water . . .*

Is thirst instinctive? If we take the claims of evolutionary psychologists seriously—if humans have instincts geared, for example, toward semantic inference and communication-pragmatics—then certainly we have a thirst instinct. Perhaps we even have a thirst module to ensure that our basic bodily need for water is never ignored. As we will see, an accurate account of thirst and all of its associated behaviors is not reducible to such simplistic concepts.

First, the basics. After decades of research it has become clear that humans, rats, and other mammals possess two distinct thirst systems: one that is triggered by eating salt and one that is triggered when we lose blood. A bartender takes advantage of the former thirst system when he serves salty peanuts and other snacks to his customers. When salt is ingested it collects in the bloodstream and, through the process of osmosis, draws water from cells throughout the body (including neurons in the brain). When cells lose water they also lose their ability to function, so we have evolved neurons that are sensitive to dehydration; thus, these neurons act as sentinels for all the other cells in the body. Through their connections with the central nervous system, these dehydration-sensitive neurons initiate a variety of physiological responses to retain water and expel salt. In addition, a neural system in the brain organizes a behavioral response, that is, drinking, that we interpret as the outward manifestation of thirst (for conceptual clarity and to avoid vague appeals to feelings, psychologists define *thirst* as "an increased tendency to drink").

A second drinking system is triggered when our vascular system detects a loss of blood. Such losses occur most dramatically when we are shot or stabbed and blood pours directly out of the system (or when we lose copious amounts of fluid as a result of vomiting or diarrhea). With a sudden loss of fluid, sensory neurons detect reduced stretch on the blood vessels, sending signals throughout the body to compensate: hormones are released and neural signals are relayed to constrict blood vessels and decrease urine formation. In addition, activation of a neural system that overlaps with the dehydration system induces the victim to seek water and drink.

Typically, thirst is induced through the combined activation of both systems, as when water evaporates from our skin and respiratory passages throughout the day and night. But both systems are needed for those times when the bodily challenge is selective and a finely tuned response is warranted. Ultimately, however, only drinking can restore fluid balance.

Motivation means very different things to different people. For many, the word conjures up images of highly paid speakers cajoling their audience to lose weight, to be happy in their marriage, or to be more productive at work. For biologically oriented psychologists, however, motivation refers to a change in the physiological state of the body that increases or decreases the likelihood that a particular behavior will be expressed. For example, imagine that lunchtime is approaching and you are walking down a busy street when, suddenly, your attention is grabbed by cheesy, meaty, doughy smells emanating from the local pizza parlor. Your head turns, then your body, and you are guided by an unseen force to walk into the restaurant, make your selection, and devour your slices. Your meal complete, you continue your daily activities and find yourself, only thirty minutes later, walking past the same pizza parlor. But this time you hardly notice. The same pizzas are being made and the same odors are wafting down the street, but you are unaware of their presence. Something has changed in you, not the pizza. We say that you have lost your motivation to eat pizza. Eating pizza can do that.

We say that eating is a motivated behavior. In addition to eating, the classic motivated behaviors are drinking, sleeping, thermoregulation, and sex. Drug-seeking in addicts falls into the same category. When we say that we are motivated to eat, we are merely referring to a measurable, predictable propensity to engage in a certain behavior. Thus, similar to our motivation to eat pizza, our propensity to stop at a water fountain to drink is directly related to the time that has elapsed since we last consumed water; we are compelled to sleep at night but rarely during the day; we are motivated to seek warmth when we are cold and seek the cold when we are hot; and abstinence from sex is never more possible than immediately after we have engaged in it.

Understanding these essential motivational processes has been a central pursuit of psychology for many years. When Wallace Craig, working nearly a century ago, observed the development of drinking behavior of young ring doves, he was surprised to find that his birds, long deprived of water, would not drink from a water bowl—even when they were standing in it! It appeared to Craig that these animals learned to perform the complete drinking repertoire only after their bill became "accidentally submerged in water," as when a dove pecked at a seed at the bottom of a water dish and reflexively swallowed the water that entered the immersed bill. Based on these and other observations, Craig conceptualized drinking and other motivated behaviors as comprising two distinct components. First, there is the appetitive component that entails seeking and approaching water; second, there is the consummatory component that entails the final act which, in this case, is drinking and swallowing water. (Note that *consummatory* is spelled with two *ms* and is not identical with *consumatory;* these two terms may overlap when we are discussing behaviors such as eating and drinking, but they do not when the behaviors are sleeping or sex.) Craig's distinction between appetitive and consummatory components of motivated behavior has influenced the thinking of many students of animal behavior, including Konrad Lorenz, but Craig's notion that the appetitive components, unlike their consummatory counterparts, must be learned has received scant attention until recently.

For their first two postnatal weeks, infant rats normally receive all of their food and water from their mother during suckling. Nonetheless, pups can exhibit non-suckling ingestive behaviors during this time. For example, by five days of age, rats that have been dehydrated with salt will compensate by drinking water, but only if the water is placed into the pup's mouth or near its snout. Thus, like Craig's immature ring doves, the consummatory response (that is, mouthing and swallowing) is expressed before the appetitive response (that is, seeking water). By fifteen days of age as weaning begins, pups begin eating solid food and drinking water. Remarkably, however, even when twenty days of age, rats still do not respond to salt dehydration by seeking water; this is remarkable because, in this study, the rats have had ample everyday experience drinking from water bottles. There appears, then, to be a bizarre disconnect between the rat's demonstrated ability to seek water—expressed throughout the day—and the consummatory response to salt dehydration, which is already in place weeks earlier. It is not until pups are thirty days of age that the connection is finally made and rats will seek water when they are made thirsty with salt.

Drinking is obviously not a recent evolutionary invention. On the contrary, it is one of the basic survival mechanisms of reptiles, amphibians, birds, and mammals. One might expect that if evolution had hardwired anything, drinking in response to dehydration would be it; yet, even in young mammals, one arm of the thirst system does not seem to know what the other is doing. Is the infant rat's inability to link one form of thirst to the need to drink a sign that this part of the thirst module has not yet been activated? If drinking is an instinct, then what are we to make of the infant rat's ineptitude?

One possible, and testable, hypothesis for the development of drinking is that it emerges gradually as a result of the pup's accumulated experiences with dry food and the related consequences of drinking water. Specifically, dry food, like salt, produces dehydration; it also produces the sensation of a dry mouth. Thus, pups might learn through experience that drinking after eating dry food relieves the discomfort of dehydration as well as the irritation of a dry mouth. Also, because the dehydrating effects of food and the restorative effects of water are subtle under normal circumstances, it may take many days for the pup to learn to respond appropriately to the food-induced state of thirst.

Ted Hall and his students examined these issues by manipulating the early experiences of rat pups. For example, in one experiment they weaned infant rats onto a liquid diet so that they were unable to experience the dehydrating effects of dry food; thus, these pups drank mother's milk until they were eighteen days of age and then were given access only to liquid food for the next seventeen days. Both liquid-reared and normally reared pups had access to water bottles, and pups in both groups regularly drank from the water bottles. Then, at thirty-five days of age, all rats were injected with a salt solution—the laboratory equivalent of eating salty peanuts—and their drinking response was measured. Consistent with the notion that dry food shapes dehydration-induced drinking, rats that were reared on the liquid diet did not increase their drinking when dehydrated. That is, unlike the rats that were reared on dry food, liquid-reared rats were unable to link their previous drinking experience with the rapid induction of dehydration.

The effects of prior experience on the development of drinking were next examined by injecting twenty-one-day-old rats either with a salty dehydrating solution or with a solution that does not produce dehydration. After twenty-four hours of access to water, all rats were maintained on a liquid diet for two more days before all were injected with the salty solution as twenty-four-day-olds; thus, at this time, half of the subjects were being dehydrated for the first time and half for the second time. It was found that only those rats that had experienced dehydration once before increased their drinking. In fact, these experienced twenty-four-day-olds exhibited appetitive responses to dehydration that were similar to those exhibited by normally reared thirty-five-day-olds. Thus, a single experience with rapid-onset dehydration is sufficient to accelerate the learning of an appetitive response.

These findings concerning dehydration-induced drinking provide stunning support for Craig's notion that appetitive drinking behaviors—the seeking and approaching of water that precedes the consummatory behavior of drinking—are learned. For such learning to recur reliably in each generation requires only that the experiences that give rise to it are reliably found in "the normal arrangement of the animal's world." Moreover, the

> finding that a specific desire as rudimentary as that for water is so strongly influenced by early experience suggests that learning may play a fundamental and formative role in appetitive and motivation systems in general, from the most biologically basic to complex human motives. Such an *acquired nature* for the form

that appetitive responses take may account for the impressive adaptability of mammalian behavior as a whole, as well as for the individual variations in adult motives, wants, needs, and cravings.

Acquired nature! What better way to defeat the nature-nurture dichotomy than to fuse two words that are normally invoked separately to sustain it?

The careful and precise thinking that characterizes this research on drinking behavior provides a stark contrast to the fanciful and unanchored theorizing of many evolutionary psychologists who conjecture wildly about a variety of vague human cognitive processes that, in some cases, cannot even be demonstrated to exist. Indeed, if a vital and ancient biological need such as drinking does not fit within the evolutionary psychologist's framework, then how seriously should we take proclamations concerning our instinctive capacity to perceive emotions, infer social relations, and read the minds of others, let alone musings concerning their neural underpinnings?

### Who's Afraid of the Big Bad Wolf?

Two men are stalking a moose. These men, however, are not hunters. Rather, they are scientists dressed in a moose suit. Approaching cautiously to within one hundred feet, one of the scientists hurls a snowball, soaked in wolf urine, at the moose. The scented snowball lands near the moose, emitting its pungent odor. Alarmed by the odor of its natural predator, the moose drops its head, retracts its ears, and erects the hairs on the nape of its neck—species-typical responses of moose anticipating a predatory attack. This moose, however, is lucky: smelly snowballs are not lethal.

Where relationships between predator and prey are well-established, prey become exquisitely attuned to the cues—odors, sights, and sounds—provided by their predators, and vice versa. In Alaska, it has been estimated that moose have been living with the threat posed by wolves and bears for nine thousand years, sufficient time (one would think) for moose to evolve specific, innate sensitivities to the odors and calls of these predators. The moose's rapid response to the urine-soaked snowball suggests that moose are indeed instinctively fearful of wolves. Perhaps moose have even evolved a wolf-module. But is the moose's fear instinctive? Thanks to our fondness for killing large animals in the wild, we may have an answer to this question.

Over the past century, the ranges of wolves and brown bears of North America and Europe have been reduced by over 95 percent. As these predators disappeared, their prey—moose, bison, and elk—enjoyed increasingly peaceful lives. In some locales, however, conservationists are reintroducing wolves and bears (or these predators are being allowed to recolonize), providing a unique opportunity to investigate the effects of predator cues on the behavior of naïve prey. Joel Berger and his colleagues took advantage of this opportunity and examined the predator-avoidance behaviors of moose in North American and European locales where predators have remained, have disappeared, and had disappeared but are being reintroduced. By tossing scented snowballs or playing audiotapes of the vocalizations of predators, Berger and his colleagues have provided us with unique information concerning the development, maintenance, and evolution of antipredator behaviors.

Moose that have experience with wolves or bears exhibit a variety of antipredator behaviors in response to the scents or vocalizations of predators. For example, moose become alert or vigilant, they exhibit aggressive responses, and they stop eating or abandon feeding sites altogether. In contrast, moose that have not had contact with predators for several generations show greatly reduced responses on all of these dimensions; specifically, they respond much less intensely to odors and vocalizations, they often fail to exhibit aggressive responses, they resume eating much more quickly, and they rarely abandon feeding sites. Thus, on every dimension analyzed, naïve moose exhibit a significant and very real vulnerability when compared with experienced moose. In fact, in Yellowstone National Park, there were no reported occurrences of predation on moose by grizzly bears between 1959 and 1992; in contrast, while grizzlies were recolonizing in a nearby area in Jackson Hole, Wyoming, ten moose were killed during the period between 1996 and 2000. As Berger and his colleagues wryly remark, these naïve moose were "conspicuously lacking in astuteness."

It is notable that such a recent disappearance of predators has resulted in such a dramatic loss of antipredator behavior. The rapidity of this change suggests that antipredator behaviors in moose arise through personal experience with predators. Support for this idea comes from repeated observations of female moose that were naïve to predators but subsequently lost offspring to wolves during recolonization. After they lost their young, these moose were significantly more vigilant when they heard wolf calls and were much more likely to abandon a feeding site; in contrast, moose that lost young for other reasons, such as starvation, did not alter their responses to wolf cues. Thus, "naïve prey have the capacity to process information about predators swiftly," even "in a single generation." As long as moose are able to survive an attack, or if attacks are focused predominantly on the young, antipredator behaviors can be learned and transmitted culturally across generations.

For evolutionary psychologists, the notion that animals, including humans, might gain crucial information about the world through individual experience is viewed as inefficient, imprecise, unworkable, and unreliable. As we have seen, however, individual experience, especially when coupled with species-typical rearing environments, is often sufficient to produce efficient, precise, workable, and reliable behavioral responses across generations. In addition, the fact that moose have so quickly forgotten the smells and sounds of their most significant predators in such a brief period of time provides yet another reason for being skeptical about the claims of evolutionary psychologists concerning human behavior.

It is important to remember that evolutionary psychologists wish to convince us that the conditions prevailing thousands of years ago during our hunter-gatherer past provided us with highly specialized modules designed to produce specific behaviors and cognitions. For some, however, our hunter-gatherer past is much too recent to account for some human instincts. Susan Mineka and Arne Öhman, for example, have investigated the human fear of snakes and have concluded that "fear and respect for reptiles is a likely core mammalian heritage" that dates back, they suggest, to the earliest mammalian experiences with dinosaurs. Given that dinosaurs became extinct at the end of the Cretaceous Period, Mineka and Öhman suggest that we are suffering from a fear hangover that has been faithfully transmitted from species to species for over sixty million years. Now that is a phobia!

Humans, of course, are not the only primates that are fearful of snakes; many other primates are as well, including rhesus macaques and squirrel monkeys. Interestingly, in contrast with wild monkeys, laboratory-reared monkeys are often not fearful of snakes. Even more odd is that while a fear of snakes is not expressed by laboratory-born squirrel monkeys that are raised on fruit and monkey chow, a profound fear of snakes (similar to that seen in wild-born monkeys) develops when they are raised on live insects. This is an odd and nonobvious path to fear that might inspire some to remain cautious when theorizing about the sources of human fear. Instead, Mineka and Öhman plow confidently ahead, borrowing from the evolutionary psychology playbook to posit the existence of a fear module that is automatically activated without any input from "more advanced human cognition." They write:

> This specialized behavioral module did not evolve primarily from survival threats provided by snakes during human evolution, but rather from the threat that reptiles have provided through mammalian evolution. Because reptiles have been associated with danger throughout evolution, it is likely that snakes

represent a prototypical stimulus for activating the fear module. However, we are not arguing that the human brain has a specialized module for automatically generating fear of snakes. Rather we propose that the blueprint for the fear module was built around the deadly threat that ancestors of snakes provided to our distant ancestors, the early mammals. During further mammalian evolution, this blueprint was modified, elaborated, and specialized for the ecological niches occupied by different species. Some mammals may even prey on snakes . . .

This last sentence is a rather weak attempt to dismiss the uncomfortable fact that foxes, raccoons, cats, and other mammals *do*, not *may*, regularly prey on snakes. It is also not clear, however, why so many humans have no fear of snakes and some—known as herpetologists—even seem to enjoy their company. What is the matter with these people? Mineka and Öhman, of course, have the answer: "Genetic variability might explain why not all individuals show fear of snakes." Indeed.

It is rather convenient for Mineka and Öhman's hypothesis that we have no means of determining whether reptiles have "been associated with danger throughout evolution" (though I relish the image of my reptile-fearing ancestors running frantically from gruesome geckos and terrifying turtles). Nor will we ever be able to test the notion that the fear module's blueprint has been evolutionarily modified, elaborated, and specialized. Still, we can ask reasonable questions.

Mineka and Öhman might be surprised to learn that snakes evolved from a non-dinosaur ancestor at least thirty-five million years *before* the extinction of the dinosaurs; thus, the notion that the human fear of snakes is a holdover from an earlier mammalian fear of dinosaurs is implausible from the start. But even if snakes evolved from dinosaurs, why would an evolutionarily ancient fear of dinosaurs get confused in our brains with a fear of snakes? Why wouldn't lizards, which look a lot more like dinosaurs than snakes, elicit the same reaction? And why do we seem to fear harmless garter snakes as much as we fear vipers and pythons? If natural selection can morph our mental image of a dinosaur into that of a snake, then why can't it provide us with the instinctive capacity to identify *dangerous* snakes? Where is our viper module?

For the sake of argument, let's concede the notion that humans have a specific fear of snakes and that snakes have played a formative role in mammalian evolution. Then why would crocodiles evoke fear as readily as snakes but not lions and tigers and bears, carnivores that likely posed at least as significant a threat to early humans (and our mammalian ancestors) as snakes and crocodiles? And how can we explain the fact that the human fear of spiders—which even has a name, arachnophobia—rivals that of snakes despite the fact that very few spider species pose a threat to us?

Common sense tells us that fear throughout the animal kingdom has many sources and is continually modified by experience and changing circumstances. Why else would birds and other animals living in protected or isolated environments, from the Galapagos Islands in the South Pacific to the Skellig Islands in the North Atlantic, show so little fear when humans approach them? It is astounding, then, that so many people, even in the face of common sense, embrace the grandiose, untestable theories of evolutionary psychologists and the static view of evolution that they endorse. Does the popularity of these theories merely reflect the fact that everyone likes a good story?

## *Mating for Life*

Here is another good story: Every newly hatched bird faces the task of identifying, affiliating, and mating with other members of its own species. For the several thousand songbird species—such as canaries, finches, sparrows, and cowbirds—the developmental task includes learning a complex species-typical song. How do these birds solve this task? Perhaps they possess innate core knowledge about their species that guides development toward the ultimate goal of mating successfully. This knowledge might be contained within an innately specified brain module. Accordingly, under the guidance of this module, birds would develop a filial imprint on their mother that, in turn, would establish a sexual imprint for their future mate. In addition, perhaps male birds possess a song-learning module that contains a mental template for acquiring their species-typical song, while females possess a similar module with a mental template for recognizing their species-typical song. Then, when their first mating season arrives, inexperienced males and females are well-prepared to do the right thing, thanks to their species-identification and song modules.

We have seen that the process by which ducks and other birds learn to affiliate with their mother and siblings cannot be so easily explained away by appeals to innate neural modules. But what about sexual imprinting, the process by which young birds develop a sexual preference for members of their own species? As mentioned in the previous chapter, Lorenz believed that sexual imprinting and filial imprinting are produced by a single process; we now know, however, that this is not the case. Thus, whatever core knowledge birds possess, it does not necessarily give rise to identical filial and sexual preferences.

One popular method for investigating sexual imprinting is to raise young birds of one species by parents of another species (such fostering experiments are made possible by the fact that many birds readily accept and raise the young of other species). The logic behind this method is straightforward: if birds prefer to mate with their own species even after being raised by foster parents of another species, then they must be exhibiting an innate attraction toward members of their own species. Such an apparent predisposition was forcefully demonstrated thirty years ago when it was shown that male zebra finches raised by Bengalese finch parents nonetheless prefer to mate with female zebra finches as adults. This finding was reinforced by subsequent experiments in which male zebra finches were raised by one zebra finch and one Bengalese finch. Despite the dual parental influences, these males grew up to exhibit a clear mating preference for females of their own species. These experiments appeared to provide incontrovertible evidence that birds possess an innate predisposition toward members of their own species that helps to ensure species-appropriate mating. Indeed, evolutionary psychologists typically rely on much less compelling evidence to argue for innate predispositions in humans.

One reason that so many students of animal behavior resist the easy appeal to innate predispositions is that there are few instances where asking the next question has not been rewarded. So what is the next question here? If a male zebra finch prefers to mate with a female zebra finch despite having extensive social interactions, from hatching onward, with adult zebra and Bengalese finches, where can we turn for an alternative explanation? In fact, the answer is buried in the previous question: What are the actual social interactions of young zebra finches with parents of two different species? Can we safely assume that each foster parent treats the young zebra finch identically?

Answering these questions required detailed observations of the behavioral interactions between young and their parents. When such observations were performed, it was discovered that the young zebra finches were not only fed most often by the adult zebra finch parent, but this parent also acted more aggressively toward it. Since this raised the possibility that social interactions influence the burgeoning sexual preferences of young zebra finches, situations were devised where the Bengalese zebra finch was encouraged to accept increasing levels of responsibility for feeding the young bird; amazingly, increasing the number and intensity of the social interactions between young zebra finches and their Bengalese parents enhanced their subsequent sexual attraction to Bengalese finches. Thus, the bias of zebra finches for other zebra finches derives from a species-typical feature of the social environment that, under normal circumstances, produces a sexual preference for their own kind.

▪ ▪ ▪

We have now become intimately familiar with the notion that a species-typical environment is up to the task of reliably shaping the development of complex behaviors across generations. But now we face an even harder challenge, that is, the situation where the normal rearing environment is always provided by the foster parents of another species. How can members of such a species find appropriate mates without having evolved a "species-identification module" to counteract the confusion of being raised by birds of a different feather?

Thirty years ago, the prolific and influential evolutionary biologist and ornithologist Ernst Mayr proposed that we envision innate and acquired traits as closed and open genetic programs, respectively. Closed genetic programs, Mayr argued, are closed "because nothing can be inserted in it through experience," whereas open programs allow for experiential input. As an example of a closed genetic program, Mayr focused on one of the most remarkable examples of evolutionary ingenuity:

> There are several groups of birds—for instance Old World cuckoos and some New World cowbirds—in which the female lays her eggs in the nest of a foster species, let us say, in the case of the cowbird, in the nest of a song sparrow or yellow warbler. The foster parents raise the young cowbird after it hatches and continue to feed it for a period of two or three weeks after it leaves the nest until the fledgling becomes independent. It then, so to speak, says goodbye to its foster parents and searches out the company of other cowbirds, which form flocks usually composed of both immatures and adults. These flocks stay together for the entire fall, winter, and the beginning of next spring. When the mating season arrives, they break up. The young cowbird mates with another member of his own species, and the female starts searching the nest of a foster species in which to place her eggs. Quite clearly, in this case the program for species

recognition—that is, the program for recognizing the appropriateness of the future mate—was contained completely in the original fertilized zygote. This is what I call a *closed* genetic program. With few exceptions, species recognition in animals is controlled by a closed genetic program.

When I first read Mayr's paper as a graduate student in the 1980s, I recall being persuaded by his argument. At that time, everyone seemed to be talking about programs, and Mayr's ideas meshed seamlessly with the increasingly popular notion that genes possessed information (the software) that was decoded by the cell's machinery (the hardware). As we have seen, this is a false dichotomy because there exists no clear distinction within the cell between hardware and software; indeed, at the time, some scientists were decrying the widespread acceptance of such simplistic and misleading metaphors. But these scientists, despite their stature, could do little to stem the tide created by the metaphorical merging of genes and computers. An idea had been born and it was intent on leading a full life.

Metaphors aside, a brood parasite like the cowbird presented a wonderful opportunity for two young researchers, Meredith West and Andrew King, to explore the mechanisms of species identification. West and King's intellectual journey has been a true scientific adventure—replete with unexpected turns and mid-course corrections—that is not yet complete. But while only small parts of this complicated and fascinating story can be recounted here, it will become clear that West and King have provided us with a model approach for unraveling the complex contributions of the social environment to the behavior of even the most "genetically closed" of all species.

Male cowbirds produce a species-typical song that is used during courtship to solicit mates; this song has been transcribed as *burble burble tsee* but is sufficiently complex that subtle modifications to the song are sufficient to alter its attractiveness to females. When West and King began rearing male cowbirds by hand in the late 1970s, they expected these birds to produce songs that matched perfectly with their catalogue of over four hundred songs recorded from wild birds from the same locale; this was what the genetically closed cowbirds were supposed to do. Surprisingly, however, the songs of the hand-reared cowbirds differed from the population norm. But then the birds did something that appeared to live up to expectations: when the acoustically abnormal songs of the hand-reared males were played back to females, these females expressed a preference—as measured by the adoption of a species-typical copulatory posture—for the *abnormal* songs. In fact, this preference was so robust that West and King, much to their chagrin today, labeled the abnormal songs as *supernormal.*

The implications, it seemed, were clear: West and King had apparently discovered "an independent system designed to insure identification during the most important context, the breeding season." In other words, they—and many others who read the report—thought they had uncovered an innate safety net. As West and King were to discover, here was an idea that meshed seamlessly—indeed, too seamlessly—with the cowbird zeitgeist. Recalling their attempts over the subsequent quarter century to reverse the impact of their early report, they regret that the "'just-so-ish' gist of the story undoubtedly acted as a magnet to hold the parts together and worked against us when we tried to pull it apart."

Why did they try to pull it apart? As West and King were soon to find out, there were several aspects of cowbird behavior that, when examined in greater detail, told a completely different story than the one they had originally believed. We can begin with one simple fact: even if hand-reared male cowbirds emit songs that females find particularly attractive, such a finding could never trump the fact that *these males are not able to mate successfully with females.* They are, to put it bluntly, duds. Moreover, the appeal of the original story is diminished further by the fact that idiosyncrasies in the methods of the 1977 report were responsible for the apparent discovery of a supernormal song; in fact, the songs of hand-reared males are not supernormal but may actually be subnormal. Finally, based on numerous studies of cowbird behavior, West and King have convincingly documented the vital importance of social interactions in the development of functional mating behaviors in cowbirds such that "the safety net is not prebuilt into either sex, but socially built up between them."

In one study, West and King captured juvenile male cowbirds, fifty to one hundred days of age, in late summer and housed them individually with either a pair of adult female cowbirds or a pair of adult female canaries (below, I refer to these males as COW-males and CAN-males, respectively). Ten months later, in early spring, these now-adult males were allowed to interact with four unfamiliar adult females—two cowbirds and two canaries. Not surprisingly, COW-males sang to the female cowbirds and courted them, although not as proficiently as males that were reared normally. The CAN-males, however, did something that was completely unexpected: they not only produced songs that contained some of the elements of canaries, but they actively courted canaries despite the presence of other cowbirds in the cage!

Next, all of the young male cowbirds were transferred to large aviaries that were filled with a rich assortment of social opportunities provided by males and females of a variety of species and ages. Under these conditions, the CAN-males continued to ignore the female cowbirds, but the COW-males now ignored them as well, focusing most of their attention on each other. As West and King describe it, the COW-males behaved like "teenagers at a dance where the boys talk together across the room from the girls." Next, when new adult male cowbirds were added to the social mix, these new adults began courting and mating with the female cowbirds while the CAN-males and COW-males continued to miss out. West and King view these and other observations as "sweeping dismissals of our original proposition of the existence of a built-in identification program geared especially to mate recognition. Under the social conditions present in the more complex settings, we saw that behavioral pieces did not suddenly or gradually snap into place. . . ." To make sure that everyone would get their point, they write humorously of their desire to title their paper reporting these new results in tabloid fashion: COWBIRD COURTS CANARY.

Like all other songbirds studied thus far—and contrary to expectations of them as so-called brood parasites—cowbirds must learn their songs through their social interactions with adults of their own species. For most songbirds, young males learn from adult males. But in cowbirds, West and King found that females play a conspicuous teaching role. What they teach is the local song dialect; for example, cowbirds from South Dakota produce a song that is distinct from that produced by cowbirds in southern Indiana and, moreover, these local dialects are transmitted stably and reliably across generations through social interactions alone. Females help to transmit this social information by providing feedback to young males as they sing to them, thereby instructing males as to the songs that will eventually prove successful for mating. This instruction comes in the form of a subtle lifting of the wing—a wing stroke—in response to song elements that match the local dialect. Here, then, is a social learning process that would make Edward Thorndike and his puzzle-box cats proud: male cowbirds begin by exhibiting an unstructured song that, through subsequent reinforcement by females, is gradually shaped into a fully functional song that has a local flair.

Given what we have learned about cowbirds (and finches), it is instructive to consider the question of *where* instincts reside. Initially, our inclination might have been to imagine that the instinctive attraction of cowbirds for other cowbirds resides somewhere within the mind (and genes) of the hatchling. But as we have seen, what brings species members together has as much to do with the behavior of adults, who preferentially interact with young cowbirds and thereby bring them into the fold. These adult preferences exist before each new generation of cowbirds is hatched; they are part of the intergenerational fabric of cowbird society. In other words, the ability of cowbirds to seek out and mate with other cowbirds, despite their isolation at hatching, is best considered a species-typical ability that emerges when species-typical social interactions are allowed to take place.

There are, then, many interconnected components that contribute to adaptive social development in cowbirds. To bring this point home, West and King ask us to consider a baseball statistic—runs batted in, or RBIs—and the fact that this statistic is typically attributed to an individual hitter. This attribution is widely accepted despite the fact that RBIs also require "a pitcher, fielders, catchers, umpires, *and* the performance of the previous batters." Given the embedded nature of RBIs within a social fabric comprising multiple interdependent parts, it is obvious that a baseball geneticist would be wasting his time searching for the RBI gene. Nonetheless, analogous searches for behavior-producing genes are regularly conducted and widely reported based on "the illusory view that the behavior we see before us is a property of the individual organism. We may be focusing at one moment in time on only this organism, but the contributions of those we are simultaneously ignoring are no less real."

West and King have given us a good story that has the added benefit of providing insight into the actual individual and social developmental processes that give rise to complex behavior. Their story cannot be summarized in a single sentence or easily explained to five-year-olds, and raises as many questions as it answers. Their story was produced through painstaking work over several decades and by refusing to ignore uncomfortable results that contradicted previous beliefs, even their own. Perhaps most importantly, their story teaches us to maintain a healthy skepticism when scientists assert, as Mayr did thirty years ago, that a behavioral program is "contained completely in the original fertilized zygote."

I like their story.

## *Food Culture*

In the early 1980s, Ran Aisner was walking through the forests of Jerusalem pine in northern Israel. Although Israel is typically associated with ancient history, these forests are quite young, being the product of a reforestation effort that began in the late 1940s. Because the previous forests in this region had been nearly eradicated a half-century earlier to supply wood for an Ottoman railway project, the strange observation that Aisner made that day could only be due to a recent arrival to the forest. The observation—stripped pine cones piled under trees—seems fairly mundane. But what flowed from this simple observation is not.

When Aisner began discussing this phenomenon with zoologist Joseph Terkel, it was not clear to them what kind of animal would be eating pine seeds in that region of Israel. Squirrels, of course, eat pine seeds, but they are not indigenous to the region. After some investigative work in the forests, the two scientists suspected that the culprits were tree-dwelling black rats. When some of these rats were captured and brought back to Terkel's laboratory, it became apparent that they were indeed capable of stripping pine cones and eating their contents; it was also apparent that they were stripping the cones in a very efficient manner, particularly surprising given that black rats had not previously been known to inhabit pine forests in Israel. Because pine cones are heavily fortified with many rows of tough scales, it was clear that the ability to remove the scales efficiently— such that the energetic benefits provided by the seeds outweighed the energetic costs of opening the cone—was no mean feat. As they observed wild rats working on the cones, Aisner and Terkel observed that each rat used a similar systematic and highly stereotyped technique, beginning at the base of the cone and removing each scale in succession with its teeth, working its way around and up the cone until all the scales had been removed so that the nutritional contents could be easily extracted and eaten.

Although rats are renowned worldwide for their ability to exploit food sources in new habitats, the young age of these Israeli pine forests ensured that the cone-stripping technique of these particular rats was a very recent innovation for this species. And if this technique, as Aisner and Terkel suspected, was being transmitted across generations, then they had perhaps discovered the rapid evolution of a novel feeding instinct—an instinct that was uniquely permitted to evolve in this isolated region because of the absence of other mammals to compete with them for access to pine cone seeds. But how could such a novel inherited behavior arise so quickly?

Rats and other mammals, including humans, transmit feeding preferences to their young through a variety of channels. When amniotic fluid and breast milk become suffused with flavors that reflect the mother's diet, fetuses and infants can develop preferences for these flavors. For example, when pregnant humans and rats eat garlic, fetuses ingest the garlic-flavored amniotic fluid and develop a preference for that flavor that is expressed shortly after birth. Some of these effects are long-lasting, influencing the foods that animals prefer to eat throughout their lifetime, thereby providing a foundation for the cultural transmission of food preferences across generations.

Flexible transmission of food preferences to young can be adaptive for omnivores, such as rats and humans, as they encounter unfamiliar foods that are safe to eat or familiar foods that have become tainted. Rats are extremely effective omnivores; indeed, rats are so difficult to eradicate using poisoned bait because a single non-fatal experience with poisoned food is sufficient to teach the rat to never again eat food near the baited area. Moreover, young rats can learn to avoid poisoned food, without experiencing that food directly, simply by smelling it on the breath of conspecifics.

Bennett Galef has devoted much of his research career to exploring the mechanisms by which rats learn which foods in the environment are safe to eat and which are not. He has shown that young rats learn about food from adults through social transmission, but this transmission does not take the form of explicit instruction or even imitation. Rather, young rats are attracted to feeding sites visited by adults. For example, simply placing an anesthetized adult rat near a feeding site increases the likelihood that young rats will eat at that site rather than at a second, identical feeding site. When adults are not present at a feeding site but their odors remain behind, young rats detect these residual odors and, as a consequence, develop a preference for the associated feeding site. Perhaps even more impressive is that young rats learn to prefer flavors when they are able to associate those flavors with other stimuli associated with other live rats, such as the sulfurous odor of rat breath. In this way, adult rats provide inadvertent information to young rats as to which foods can be safely eaten. Finally, lest we be tempted to postulate a hardwired instinct in young rats that directs them to *listen* to what the adults are *telling* them, it is important to point out that pups reared artificially (such that they do not interact normally with their mother and siblings) are not attracted to feeding sites marked with the odors of adults. Thus, normal developmental experiences are necessary for this food-learning capability to arise, a theme that by now should be quite familiar.

Galef's extensive work has provided invaluable insights into the mechanisms by which young rats learn *what* to eat. With the Israeli black rats, however, Terkel and his students needed to determine the mechanisms by which rats learn *how* to eat. Many possibilities were tested and excluded. For example, in one experiment it was shown that when 222 wild rats raised in the laboratory were given pine cones (and only pine cones) to eat for several weeks, only 6 of them learned to strip the cones in an efficient manner. This is an extremely low rate of success, but even this low rate could be sufficient to get the stripping technique started in a population. Disseminating the technique to others, however, was apparently not accomplished through observational learning: when Aisner and Terkel housed stripping and non-stripping rats together to see whether a non-stripper could somehow learn from the stripper, none of the non-strippers learned the technique.

If adults were not transmitting the stripping technique to each other, perhaps they were passing it down to their offspring. This possibility seemed likely when it was found that thirty-one of thirty-three pups born to stripper mothers were stripping cones on their own within three months of age. But instead of proclaiming the discovery of an innate cone-stripping skill or rushing to search for a cone-stripping gene, Aisner and Terkel performed a standard cross-fostering experiment to determine the source of the infants' *knowledge*. Accordingly, pups of stripping mothers were either raised normally or cross-fostered to non-stripping mothers, and pups of non-stripping mothers were either raised normally or cross-fostered to stripping mothers. The results were unequivocal: all pups exhibited stripping capabilities in accordance with the stripping abilities of the mother that raised them and not necessarily the mother that gave birth to them. Clearly, pine-cone stripping is a socially transmitted skill. It is not innate. There is no gene to discover.

Numerous other experiments were devoted to revealing how this skill is transmitted. When the experimenters opened cones, thereby exposing the seeds inside, and presented the cones to pups, the pups extracted the seeds and ate them; nonetheless, this experience provided no impetus for them to learn how to open pine cones. But when pups were exposed to cones in which some scales had already been removed, most pups were able to continue the stripping process to completion. Because mothers are the most likely source of partially opened cones, it was concluded that "black rat pups acquire the pine cone stripping technique by snatching cones that the mother has already partly opened." The mother, therefore, is "the key factor in the pups' learning process and enable[s] this fascinating feeding behavior to be culturally transmitted from one generation to the next."

Galef has credited Terkel and his students with having given us "a better understanding of the origins of behavioral traditions in free-living rats than in any other nonhuman, mammalian species." Ironically, as the study of animal traditions is just now beginning to find its footing, John Tooby, Leda Cosmides, and other evolutionary psychologists wish to deemphasize our own impressive capacity to transmit cultural information quickly, reliably, and efficiently. For them, it is our evolutionary heritage that has played the primary role in shaping our contemporary capacities and tendencies to think and act as we do, with culture playing a subservient role. At this time, however, we simply do not know which features of our psychology and our culture have been shaped by evolution, or how biological and cultural forces shape our psychology during development. One can, in other words, accept the validity of the questions posed by evolutionary psychologists without blindly subscribing to their preferred modes of explanation. For example, Tooby and Cosmides argue that

> the human psychological architecture contains many evolved mechanisms that are specialized for solving evolutionarily long-enduring adaptive problems and that these mechanisms have content-specialized representational formats, procedures, cues, and so on. These richly content-sensitive evolved mechanisms tend to impose certain types of content and conceptual organization on human mental life and, hence, strongly shape the nature of human social life and what is culturally transmitted across generations.

But even rats, cowbirds, and moose do not fit Tooby's and Cosmides's assumptions of how adaptive problems must be solved. Moreover, their bold pronouncements float unanchored without substantive empirical support (other than question-and-answer surveys of human behavior and experiments whose results are routinely contradicted by subsequent research). All too often, the pronouncements of evolutionary psychologists do not jibe with what we actually know about behavior in humans and other animals. For example, when Steven Pinker composed his list of human instincts and included on that list *knowledge of what food is good to eat,* one can only imagine that he simply assumed that such knowledge must be instinctive.

It is a sad fact that most studies of human species-typical behavior carried out thus far do not (and in some cases cannot) come close to approaching the empirical rigor demonstrated in the examples of non-human animal research already described. Thus, the limitations of studying human instincts and their origins should temper the enthusiasm of those who wish to draw grand conclusions about the influence of our evolutionary past on contemporary human behavior. I am not in any way denying the validity of asking evolutionary questions about human behavior because there can be no doubt that we have much to learn by pursuing this line of inquiry. On the other hand, the McDougallization of this problem by Tooby, Cosmides, Pinker, and so many others is unlikely to get us closer to the true understanding of human nature that we all desire.

### *Talking Points*

Communication through language is a species-typical behavior of humans. *Clearly.*

It is universally expressed under normal rearing conditions. *True.*

It is an instinct. *Okay, if it makes you happy to say so.*

"But if there is a language instinct, it has to be embodied somewhere in the brain, and those brain circuits must have been prepared for their role by the genes that built them."

This last statement, from *The Language Instinct* by Steven Pinker, is truly unfortunate. First, Pinker's leap from language to genes is voluntary, not, as he implies, mandatory. Second, as we have seen repeatedly, there is no meaningful sense in which an instinct can be said to be embodied completely within the brain; if birdsong cannot be pigeonholed that easily, then how can human language? And finally, as described in Chapter 4, it is simply false—on a number of levels—to state that brain circuits are built by genes. Genes do not *build* anything.

Nowhere is the marriage of noble aims and naïve explanation more evident than in the attempts of Pinker and his mentor Noam Chomsky to reduce the vast complexities of human language to innate, hardwired, genetically determined, neural modules. Chomsky's contributions to linguistic theory are, of course, prodigious. He was the first to uncover the underlying grammatical structure of human language, a structure he called Universal Grammar (UG). In the wake of B. F. Skinner's contention that language, like any other behavior, is learned according to operant principles, Chomsky countered by arguing that languages are unlearnable in the Skinnerian sense. The only possible explanation, he argued, was that children, for whom language learning comes easiest, possess an innate language competence. Eventually, in the tradition of contemporary nativism, this competence was melded with the concept of a neural module.

Discovering universal features among the diversity of human languages was Chomsky's seminal contribution to linguistics, but does it necessarily follow that UG is innate? Imagine a terrestrial disaster that has killed every human on the planet except those who happen to speak Japanese. In this new Japanese-only world, the grammatical rules specific to Japanese would now be, by definition, universal. Indeed, Chomsky's UG would no longer be detectable because there would no longer be other languages with which to compare Japanese. Using Chomsky's logic, one could now argue that the ability to speak Japanese is an innate competence which, from our pre-catastrophe perspective, we know not to be the case.

UG reflects the underlying structure of language because all human languages had a single, common ancestor, and because the genealogical thread that links this common ancestor to all current languages has never been broken. But given that UG has been faithfully transmitted from generation to generation, and given that children are such able language learners, doesn't this support Chomsky's view that UG evolved because children innately possess a language acquisition device that decodes for them the inner workings of human language? As Terrence Deacon explains in *The Symbolic Species*, however, there exists a more biologically plausible option:

> I think Chomsky and his followers have articulated a central conundrum about language learning, but they offer an answer that inverts cause and effect. They assert that the source of prior support for language acquisition must originate from *inside* the brain, on the unstated assumption that there is no other possible source. But there is another alternative: that the extra support for language learning is vested neither in the brain of the child nor in the brains of parents or teachers, but outside brains, in language itself.

In technical terms, we say that humans and languages coevolved.

In Chapter 2, I described how human eating patterns and eating utensils coevolved, with changes in human behavior driving the need for new utensils that, in turn, made new behaviors possible. Behaviors produced in this way even can attain a cultural permanence that persists even after its connection with its historical past is severed. For example, recall that the American crisscross method of eating began with a curious reliance on spoons at the colonial dining table and has outlasted the spoon's replacement by the fork. Creatures from another planet (perhaps in collaboration with evolutionary psychologists from our own) might hypothesize that the different eating styles of Americans and Europeans—coupled as they are with the identical arrangements of their silverware—are caused by a geographically specific American instinct governed by a dedicated brain circuit that is built by a "crisscross gene." Of course, a simple cross-fostering experiment in which children are exchanged between the two continents would quickly falsify such a hypothesis.

As with dining habits, language also has a cultural permanence, whether because it has been written down or because it has been transmitted as part of a people's oral tradition. But which aspects of a language will be retained and which discarded? Deacon argues that those aspects of language that have been retained are exactly those that became most easily learned by children. Accordingly, children are such efficient language learners because, unbeknownst to them, the linguistic environment into which they are born has been custom-fitted to the peculiar way that their minds work. In Deacon's words, "Children's minds need not innately embody language structures, if languages embody the predispositions of children's minds!"

Deacon cleverly supports his thesis concerning language and perceptual biases with an analysis of the evolution of color terms. It has been noted in cross-cultural studies that languages differ as to how many colors have been given distinct names. The languages with the fewest color terms have words for black, white, and red. If another color term was included, that color is always green; next blue, yellow, or both. Is this spooky cross-cultural similarity only explicable by postulating the existence of a color-term module? Of course not!

Deacon reviews well-established evidence that the prototypes of red, green, yellow, and blue across many different languages match with impressive precision the actual processing of these colors by the retina and the visual portions of the brain. For example, neurophysiologists have shown that red and green are processed as pairs by the visual system, as are blue and yellow (the pairs are called complementary colors); indeed, the dominant form of human color blindness impairs the discrimination of red and green, while the less dominant form impairs the discrimination of blue and yellow. Thus, Deacon argues that linguistic references to colors have come to match the perceptual biases of the nervous system, a case of a "neurological bias acting as a relentless force in social evolution."

Thinking in terms of innate modules for language is like a crutch that can only be used in the mud: it may keep us upright but it does not help us get anywhere. Once freed from the influence of modules, the complex interactive dynamics of language learning and language evolution are evident once again. To explain this complexity, Deacon focuses on such issues as brain size and the peculiarities of the human vocal tract, but there are undoubtedly many contributors to our uniquely human capacity for language that have not yet been identified. Clearly, this journey has only just begun.

Historically, the instinct concept has been employed as a solution to the problem of how to explain the exquisite fit between organism and environment. In the struggle to explain this fit, we have bequeathed immense power to this single concept so that, like a black hole, it has devoured every related concept in its neighborhood. As noted in Chapter 1, Patrick Bateson counts at least nine scientific meanings for instinct, including innate, not learned, unchanged once developed, shared by all members of the species, controlled by a distinct neural module, genetically determined, and adapted during evolution. As should now be clear, Bateson is correct when he writes that one meaning of instinct "does not necessarily imply another even though people often assume, without evidence, that it does." The only way out of such confusion is through the accumulation of evidence by carefully and rigorously dissecting actual behaviors during development.

By now, it should be apparent that the term *instinct* is merely a convenience for referring to a variety of complex behaviors whose origins we do not fully understand. Thus, the more we learn about a behavior, the less compelling the justification for describing it as an instinct. When we discover that pecking in gulls toward its mother's beak emerges from a subtle perceptual bias or that ducklings must have heard their own embryonic vocalizations in order to develop a filial preference, we have gained an insight that dissolves our concern about what instinct really means. Similarly, when we realize that fetuses learn, our obsession with what is inborn dissipates. When rats must learn something as basic as how to approach water to drink or when moose lose their fear of wolves when these predators are no longer around to attack them, the evolutionary psychologist's fixation on hardwired neural modules seems naïve. And when we observe the efficient and elegant transmission of cultural information in cowbirds and black rats, we appreciate that there is more to heredity than genes.

All complex behaviors are composed of sub-behaviors, each of which is induced at each stage of development, often in nonobvious ways. Thus, DNA, cells, behavior, and our physical, social, and cultural environments interact continuously and dynamically, in real time, to create the behaviors that have revealed themselves to many as the products of divine or genetic design. Evolution, of course, plays a major role in the emergence of complex behavior, not by focusing its attention on genes, but by selecting for the entire developmental manifold. Natural selection simply does not care *how* the final outcome is achieved; but we should care, especially if our aim is to understand how we came to be who we are and how we can improve our lives when development goes awry.

Having surveyed some of the best that animal research has to offer to our understanding of instinct, it is now time to turn our full attention to humans. The issue that will now be tackled—the growing attraction of developmental psychologists to nativism—is rarely analyzed within the context of animal behavior. This is unfortunate because, while there are many issues to be addressed that are unique to developmental psychology, some critical elements are the same. For example, human infants, like other animals, do not use language to communicate. The people who study them do.

# 7

# THE NATIVISTS ARE RESTLESS

NATIVISTS BELIEVE THAT HUMAN INFANTS are born with the ability to use reason to make sense of a world that they have not yet directly perceived. This reasoning ability, according to nativists, acts upon core knowledge that is pre-built into the infant's brain and that reflects a specific understanding of how the real world works. Using experimental techniques specially designed to reveal this infant knowledge, nativists claim that infants, for example, know that two solid objects cannot occupy simultaneously the same location in space, that objects do not change their motion without a cause, and that numbers exist. And now that the core of human knowledge has purportedly been revealed, it is claimed that this innate knowledge provides the foundation upon which more mature and elaborate human knowledge and concepts are built. Thus, nativists argue that knowledge expands from the inside out and draw an unambiguous line between what is innate and what is learned. This is the nativist creed.

As we have seen, one of the common scientific meanings of *instinct* is "present at birth," that is, "innate." Thus, the nativist perspective is the intellectual relative of disciplines for which instinct has been a central theme, including ethology and evolutionary psychology. Occasionally, the connections between nativism and these intellectual relatives are crystal clear. For example, when Steven Pinker presents his list of "modules, or families of instincts," his first entry concerns the nativist claim that human infants possess an intuitive understanding of physics. Similarly, when developmental psychologists Elizabeth Spelke and Elissa Newport rationalize the infant's innate capacity for navigation, they invoke the standard argument of evolutionary psychologists. "In the hunter-gatherer societies that existed throughout most of human evolution," they write, our ancestors "could ill afford to become lost during their long travels from home in search of food."

In contrast, it is irrelevant to many students of animal behavior whether instincts indicate actual knowledge about the world as long as animals behave as *if* they possess that knowledge. In fact, as we saw in the previous chapter, many instinctive behaviors do not reflect knowledge but rather emerge from subtle perceptual biases. We have also seen, for example, that only a few hours of experience are sufficient to alter a chick's pecking behavior dramatically. Nativists, however, argue that infants possess actual knowledge about the real world, that they can apply reason to that knowledge, and that these capacities go far beyond mere perceptual biases. That almost all of their claims are based on experiments using infants that have already amassed at least three months of postnatal experience is considered irrelevant to their claims concerning *inborn* capacities. I will repeat this last point because it is so important. Nativists claim that our core knowledge is innate—inborn—based predominantly on studies using infants that have experienced over three months, or two thousand hours, of postnatal life. Apparently, nativists believe that human gestation is twelve months long.

It would be naïve to expect that the lessons learned in the previous chapter would shake the nativist's commitment to the dichotomy between what is innate and what is learned; or their fixation on birth as the defining moment in development; or their disregard for the notion that three months of postnatal experience (let alone many months of prenatal experience) might be relevant to the issues that they address. Nonetheless, it is an unfortunate reality of contemporary science that once a theoretical claim has gained momentum, regardless of how fantastic that claim may be, the burden shifts to the naysayers to provide evidence against that claim. After years of silence and submission, such evidence is rapidly accumulating.

As is so often the case with fantastic scientific claims, there are two questions that need to be addressed. First, does the evidence support the claim? And second, what is it about the claim that makes it so popular and so resistant to negative evidence—so alluring? Regarding the latter question, there is little doubt that nativists have the easier sell. Many people *want* to believe that babies are brilliant and they find it *easy* to understand nativist claims. In addition, human infants—like chicks, cats, moose, and monkeys—cannot talk, a limitation that redounds to the benefit of those who prefer rich interpretations of infant cognitive capabilities. Why? Because when their subjects cannot talk, researchers are free to interpret nonverbal behaviors as they wish, secure in the knowledge that they are unlikely to be corrected when they go too far with their interpretations.

Fortunately, psychologists have learned important lessons over the years concerning the pitfalls of overinterpreting the behavior of subjects who cannot speak. Unfortunately, these lessons are often forgotten.

### *Muted Minds*

Anyone who has read a book or watched a television program about animal behavior has likely come across the story of Clever Hans, the horse owned by an elderly Viennese man named Wilhelm von Osten. Clever Hans was seemingly capable of answering complex questions posed to him by tapping his right hoof on the ground (when asked mathematical questions) or shaking his head (when asked yes/no questions). Clever Hans only responded to German questioners and performed best in the presence of von Osten, although others humans were eventually able to communicate effectively with the horse.

By 1904, Hans had achieved such notoriety in Berlin and elsewhere that a group of eminent scientists formed the September Commission to investigate the horse's prowess. Because the study of animal behavior was still in its infancy, the members of the commission did not really know how to critically test the horse's purported skills. Although members of the commission were aware that they may have missed hidden cues and tricks, they nonetheless completed the Commission's assignment by concluding that Hans was indeed answering the questions posed to him without human guidance.

In the end it fell to a young graduate student, Oskar Pfungst, to reveal that the horse, although indeed clever, was being cued ever-so-subtly and unknowingly by von Osten and those others with whom the horse became familiar. In a critical experiment, Pfungst examined the horse's performance under conditions in which Hans was asked a question and von Osten was blind to the question, and then in which von Osten was asked a question and Hans was blind to the question. This experimental manipulation provided a clear result: only when von Osten knew the question was the horse able to provide the answer. The conclusion was revolutionary: humans can add!

I have not recounted the story of Clever Hans because I think that nativists are inadvertently providing cues to human infants during their experiments. No, the story of Clever Hans is bigger than that. This story retains its interest for students of behavior because it is the iconic example of the benefits of caution and skepticism in the face of fantastic scientific claims. Moreover, it illustrates the traps that are set for us as we investigate minds that cannot communicate through language. And it also reminds us that our interpretations of behavior—horse or human— are easily distorted as they pass through the lens of our expectations.

▪  ▪  ▪

The story of Clever Hans is funny. Autism is not. And yet there is a connection between the two that, once again, helps us to recognize the difference between sound and unsound psychological thinking. The lesson to be imparted here concerns the need to use critical and realistic thinking when considering what the mind is and how it develops.

Stephen Hawking is one of the great physicists of the twentieth century. He has also, because of the devastating motor impairments brought on by amyotrophic lateral sclerosis (ALS), lost the ability to communicate using his voice. Indeed, because of his extraordinary fame, many of us associate Hawking's voice with the clipped, metallic-sounding device that he uses to communicate. This computer-generated device is controlled by small movements of his fingers, his last remaining verbal connection with the outside world. The computer, therefore, facilitates Hawking's ability to communicate, just as similar devices are now routinely used to facilitate communication in paraplegics and other individuals with neurological diseases and insults.

Stephen Hawking is an adult and that is the nub of the issue that we are addressing here. As an adult, and before the onset of his disease, Hawking had learned to use language to communicate. He had formed adult relationships, had engaged in adult conversations, and had developed adult views of the world around him. The effect of ALS on his motor abilities had a relatively small effect on his sensory abilities and his ability to think and construct sentences. His ultimate fate—complete motoric isolation from the world—does not negate the lingering presence of a functioning mind. In this case, as we have observed the wasting of Hawking's motor skills and the increasingly clever compensatory mechanisms available to him for communication, we are confident that a brilliant mind persists regardless of the ease with which we are able to communicate with it.

Thus, before the onset of his disease, Hawking repeatedly demonstrated his competence to communicate with others but then, with the onset of his disease, his ability to perform his communication skills diminished independently. But what are we to make of a mind that, like Hawking's, appears distant and isolated but, unlike Hawking's, has never engaged in adult, adolescent, or even child-like interactions with the world around it? This is, of course, the situation that exists with human infants. Indeed, nativists claim that evolution has provided human infants with the *competence* to understand certain aspects of physics and mathematics but has not provided them with the concomitant capacity to express—or *perform*—that competence. Analogously, it has recently been claimed that children with severe autism and other developmental disabilities are competent to communicate normally but are unable to express that competence without the assistance of others. This last claim is worth examining in detail.

I was first introduced to an approach for opening the minds of severely autistic children while watching television one evening in 1992. In the newsmagazine *PrimeTime Live*, a miracle was being revealed before the American public: an autistic child, typing on a keyboard, was communicating a rich mental life heretofore unimaginable. By the child's side was an adult facilitator who helped the child to type by holding her wrist. As Diane Sawyer gushed over the poignant comments emanating from the tip of this child's finger, the facilitator quietly and dutifully steadied the child's arm. This apparent breakthrough, called Facilitated Communication (FC), promised a reinterpretation of autism (and other developmental disorders) as a problem of the motor system in which cognitive functioning remains intact.

FC began in Australia when a special education teacher, Rosemary Crossley, came to believe that many of her disabled children possessed richer mental lives and capacities than their motor disabilities allowed them to express. These capacities, Crossley decided, could be revealed by providing hand or arm support to children to aid them in their struggle to communicate. This method came to the attention of Douglas Biklen, an American psychologist, who quickly promoted FC as part of a wider effort to justify the integration of children with developmental disabilities back into the general school environment. In a report published in the *Harvard Educational Review* in 1990, Biklen provided testimonials from autistic children (using facilitators) who, despite no previous indication of an ability to communicate, nonetheless expressed sophisticated opinions concerning their plight as stigmatized individuals in our society. In 1992, Biklen established the Facilitated Communication Institute at Syracuse University which, according to their Web site, promotes the FC method "for individuals with severe disabilities, including persons with labels of mental retardation, autism, Down syndrome and other developmental disabilities."

In 1992, I knew very little about autism and had never heard of FC. But by the end of Diane Sawyer's report I was literally jumping up and down in anger. I was not angry because I am opposed to new and innovative techniques for improving the lives of the developmentally disabled— of course not. But I was shocked and appalled that such an obvious scam—whether perpetrated intentionally or unintentionally—was being cruelly promoted to a national audience.

How, you ask, could I possibly be certain that FC was a scam based on a puff piece on a network television show? My answer is simple: the report depicted children punching out sentences with one index finger *without looking at the keyboard*. The task seemed extraordinary even to Sawyer, who asked one severely autistic teenage boy—who was "typing" in a standing position with one arm firmly grasped by the facilitator and his other arm flailing uncontrollably—how he could type so accurately with his eyes completely averted from the keyboard. The boy, Sawyer relates to us, responded that he "can see the letters in his mind and can aim his fingers at an imaginary keyboard." Satisfied, Sawyer continued with her story.

In fact, this is where her story should have ended. Although you cannot see me writing these words at my computer, I can assure you that I, like most people who spend too much time at the computer keyboard, am a proficient typist. And here is how I type *autism* when I punch out the letters with a single, unanchored finger while looking away from the keyboard: XSJGJC. Try it—it is impossible.

The real claim of FC proponents, however, is that autistic children are best able to communicate using keyboards when provided with steadying support from a facilitator. The question that then arises is whether the facilitator provides more than mere support. Although the facilitators themselves may be convinced that it is the child, and the child alone, who is typing and communicating, there is a simple way to test this underlying assumption that is identical to that used by Pfungst to unravel the mystery of Clever Hans. Specifically, we can direct questions to the autistic child and facilitator independently and determine who must receive the question in order for a correct answer to be given. When such experiments have been conducted, the results have been unequivocal: correct answers are given in accordance with the knowledge of the facilitator, not the child. In short, FC is a sham.

As with ESP, homeopathic medicine, and other similar movements, no amount of negative evidence or instances of fraud will extinguish the FC movement. But extinguishing that movement is not my goal here. Rather, what I wish to emphasize once again is that *naïve theories of the mind and behavior make us vulnerable to fantastic claims concerning human psychological phenomena.*

Consider the basic assumptions underlying the FC movement, namely, that children who have never uttered a word, who have never exhibited any communicative skills through gesturing, and who have received little or no instruction in basic language skills such as reading and writing, can nonetheless covertly develop such skills to the point where they can spontaneously communicate through a facilitator at a level that meets or exceeds our expectations for a normal child of that age. Biklen, faced with the obvious problems with these assumptions, has suggested that these children must be learning from passive exposure to the environment (such as learning from television and watching their brothers and sisters do their homework); such a suggestion, coming from an educator of all people, is extraordinary.

Just as Biklen claims that FC enables autistic children to bypass a motor handicap, nativists claim that their techniques allow infants to bypass a language handicap. But there is one important difference between the two claims: whereas FC was easily shown to be bogus using a blind experimental procedure in which questions were posed selectively to the child or the facilitator, the edifice erected by nativists, shaky as it may be, cannot be toppled by any single experiment. Thus, in our evaluation of nativism we must take a different tack. First, we must critically examine the experimental methods that nativists use to generate their data. Then, we must determine whether the bulk of the experimental evidence actually supports the nativists' preferred interpretations of their data.

## *Leading Questions*

The most direct method for establishing what an individual knows is to ask a question. For example, I might ask, "Who was the first president of the United States?" If you answer "George Washington," there is no ambiguity and no need for complex interpretation. The answer says it all.

George Washington is a figure in history. He does not possess the timelessness of mathematics, geometry, or physics. Thus, we would never expect humans to have inborn intuitions about George Washington. We might, however, expect such intuitions about numbers, lines, and gravity, especially if we are nativists.

But how can we demonstrate the reality of innate intuitions? Is it sufficient to simply ask questions? The Socratic method, practiced by Socrates and preserved in written form by his student, Plato, is a highly evolved form of the question-and-answer approach to revealing the origins of knowledge. And in the *Meno,* one of Plato's most popular dialogues, the Socratic method is applied by the master to the question of geometric knowledge. Indeed, according to Elizabeth Spelke and Elissa Newport, Socrates may have conducted "the first study of the development of geometrical knowledge."

Although the *Meno* begins with an inquiry into the origins of virtue, it is most famous for a passage in which Socrates probes the mind of a young slave boy to demonstrate the thesis that learning is merely a form of recollection. The demonstration begins with the premise, accepted by the major players in the dialogue, that the boy has had no formal training in geometry. Then, the geometrical problem is posed: Given a square with sides of a particular length, by how much would these sides need to be increased in order to double the area of the square? The exchange with the boy begins with Socrates drawing squares in the sand and asking, "Now boy, you know that a square is a figure like this?"

Boy: Yes.

Socrates: It has all these four sides equal?

Boy: Yes.

Socrates: And these lines which go through the middle of it are also equal?

Boy: Yes.

Socrates: Such a figure could be either larger or smaller, could it not?

Boy: Yes.

The drawing in the sand becomes increasingly complicated as Socrates guides the boy toward the final solution. In the process, the boy answers Socrates's rather simple questions and, on occasion, he even answers incorrectly. The exchange ends in this way, with Socrates again pointing to a figure in the sand:

Socrates: How big is this figure then?

Boy: Eight feet.

Socrates: On what base?

Boy: This one.

Socrates: The line which goes from corner to corner of the square of four feet?

Boy: Yes.

Socrates: The technical name for it is "diagonal"; so if we use that name, it is your personal opinion that the square on the diagonal of the original square is double its area.

Boy: That is so, Socrates.

Socrates: What do you think, Meno? Has he answered with any opinions that were not his own?

Meno: No, they were all his.

Having extracted this confession from Meno, a cascade of rapid-fire conclusions ensues, such as "these opinions were somewhere in him"; "knowledge will not come from teaching but from questioning"; "the spontaneous recovery of knowledge that is in him is recollection"; "no one ever taught him"; "if he did not acquire [opinions] in this life, isn't it immediately clear that he possessed them during some other period?"; "when he was not in human shape?"; "may we say that his soul has been for ever in a state of knowledge?"; if so, then "the soul must be immortal." At this point, the dialogue resumes its discussion of virtue, ultimately concluding that "virtue will be acquired neither by nature nor by teaching." What possibility remains? Socrates concludes that virtue is inserted into the soul by "divine dispensation."

Clearly, the *Meno* is a nativist text of a different sort. Our aim here, however, is not to explore the intricacies of Socrates's views concerning the immortal soul. Rather, we are interested in the feasibility of exploring the origins of human knowledge through the use of the Socratic method or any similar approach that relies on questions and answers (and thus language). In other words, did Socrates succeed in his aim of extracting knowledge from the boy that was already "somewhere in him?" Clearly not. As Jacob Klein notes in his commentary on the Meno, "the direction that the inquiry takes is completely determined by the order of the questions that Socrates asks" and "it is

Socrates who draws all the figures and, above all, the diagonals on which the solution of the problem entirely depends." Bertrand Russell puts it more colorfully. Socrates, he writes, asks "leading questions which any judge would disallow."

During a trial, one lawyer may object to a question on the grounds that opposing counsel is leading the witness. For example, on direct examination, a question like "When you arrived at eight o'clock, what did you see?" would elicit an objection from opposing counsel on the grounds that it is a leading question, that is, it presumes a fact—arrival time—that has not yet been established. In other words, the lawyer is putting words into the mouth of the witness. Leading questions are permitted, however, during cross-examination, in part because it is presumed that an antagonistic witness would resist the attempts of an opposing lawyer to put words in his mouth. There is, then, a role for leading questions, but that role is restricted to examination of an issue *after* the facts have been established. Similarly, Russell noted that the Socratic method, resting as it does on leading questions, may help to illuminate problems of fact and logic, but "it is quite unavailing when the object is to discover new facts."

In our legal system, it is understood that the ability to communicate —to use language—does not guarantee competence as a witness. For example, when children are presented as witnesses, they must first be examined—in the presence of a judge—to assess their capacity to understand the questions being asked of them. It is not surprising, then, that children who cannot communicate with language at all are not deemed fit to act as witnesses in a trial. But this does not necessarily mean that preverbal children, and even infants, are not able to tell us what they know. Infants may not talk, but they are able to move their legs, arms, mouth, tongue, and eyes, and developmental psychologists have exploited these bodily movements to explore the infant mind. Nativists interpret their experiments as evidence that infants possess innate core knowledge and specialized capacities.

Others object, however, on the grounds that nativists are leading their witnesses.

You be the judge.

### *Pale Imitation*

Let's begin at birth.

A human infant cannot do very much when it is born. Sleeping and suckling are its two primary occupations; it cannot walk or even crawl; lifting its head is a chore and balancing its head is just as difficult. But if you place an infant on your lap, support its head with your hands, and make funny faces, like sticking out your tongue or opening your mouth, it has been claimed that even a newborn will imitate you by also sticking out its tongue or opening its mouth. This startling claim was made by Andrew Meltzoff and Keith Moore in a paper, published in *Science* in 1977, describing two experiments that seemed to establish imitative abilities in infants only several weeks old. Subsequent experiments extended this imitative capacity to within minutes of birth, thereby ruling out a role for learning and laying the foundation for the claim that imitation is innate. Many developmental psychologists were stunned by these findings because they suggested an ability that goes far beyond what was thought possible at such a young age. Indeed, the revered Swiss developmental psychologist Jean Piaget had asserted many years before that imitation does not appear before the age of one year.

Imitation is a simple word that connotes a complex process. Imagine that you are a newborn infant who is capable of imitation: you see a face hovering over you and a big red tongue emerging from it (of course, newborns do not see "faces" and "tongues" but rather only shapes and colors). The images of face and tongue are detected by your retinas and processed by your brain. Once these sensations have been processed, the innate urge to imitate triggers a behavioral response. To do this, however, requires that you instinctively produce a facial or bodily expression that mirrors the stimulus above you—the tricky part is that you don't have a mirror. Even more astounding is that you do this without any experience. In other words, you are innately hardwired such that seeing a protruding tongue automatically activates your tongue muscles, seeing an open mouth automatically activates your jaw muscles, and seeing moving fingers automatically activates the muscles that move your fingers.

Even if human newborns possess an innate capacity for imitation, why would evolution go to so much trouble to ensure that our brains possessed the neural wiring and urge to imitate the behavior of others at birth? Meltzoff thinks he knows why. Imitation, he writes, "holds the key to our understanding what it is for others to be like us and for us to be like them." In other words, imitation lays the foundation for empathy as well as the development of a theory of mind, that is, a theory that allows us to understand the goals and intentions of others and, therefore, predict future behavior. Possession of these high-level abilities, it is argued, confers obvious evolutionary benefits. Wrapping all of this into one neat package, Meltzoff writes:

> Metaphorically, we can say that nature endows humans with the tools to solve the "other minds" problem by providing newborns with an imitative brain. In ontogeny, infant imitation is the seed and the adult theory of mind is the fruit.

It would be negligent of me not to point out that many hard-nosed psychologists roll their eyes when their colleagues make bold pronouncements about goals, intentions, empathy, and theory of mind. None of these terms and concepts is particularly well-defined which, one might think, would give pause to those committed to using them. Unfortunately, the attraction of psychologists to vague mentalistic concepts is enduring. But we will not dwell on this issue here. Instead, we will focus on a much simpler question: Do human infants truly possess an inborn capacity for imitation?

One might think that a "phenomenon" discovered over twenty-five years ago and that is now used as the conceptual foundation for a theory of how we develop our capacity for empathy would be beyond reproach. Indeed, as recently recounted by Meltzoff, the basic imitation finding has been "replicated and extended in more than two dozen studies from thirteen independent laboratories." According to him, there is no dispute, no controversy, no room for doubt. Indeed, it is true that most developmental psychologists now have an unquestioning belief in this extraordinary phenomenon. But contrast this state of affairs with the recent opposing claim that "there is little basis for the hypothesis that neonates can imitate oral gestures. The widespread acceptance of [Meltzoff's] hypothesis may be due in part to its dramatic appeal and to the predilection to attribute innate competencies to babies." How can we possibly reconcile such diametrically opposing views?

Many developmental psychologists were skeptical of the imitation claims from the start. This skepticism was based partly on the very nature of the claims and partly on the dubious methods that were used in the original experiments. Moreover, even in the hands of Meltzoff and Moore, imitation was surprisingly difficult to demonstrate because infants spontaneously exhibit a variety of behaviors, including tongue protrusions, mouth openings, and movements of the head, arms, and fingers. Consequently, at best, imitation is a subtle phenomenon whose demonstration requires many testing trials and complex statistical argument. For a behavior that is supposed to play such a seminal role in mental development, it is curious that so much effort is needed to coax it out of the data.

We begin our analysis of the imitation phenomenon by noting that Meltzoff and Moore's original study detailed infants' abilities to imitate four behaviors: lip protrusions, mouth opening, tongue protrusions, and finger movements. Over the ensuing years, Meltzoff and other researchers restricted their investigations to a narrower set of movements, typically comprising tongue protrusion and one other behavior; most often, the second imitated behavior was mouth opening. In a study published in 2001, Moshe Anisfeld examined all of the imitation studies conducted by Meltzoff and others and showed that only tongue protrusions by adults reliably elicit tongue protrusions by infants. Indeed, most attempts to demonstrate imitation in infants have produced negative results. In response, the proponents of infant imitation try to explain away negative findings by arguing that insufficiently sensitive methods were used. There is, however, one fundamental problem with this argument. As Anisfeld has shown in his survey of previous research, investigators who have successfully replicated the tongue protrusion effect are typically not successful using other behaviors *in the same experiment*. For example, Anisfeld and his colleagues recently showed that newborns produce more tongue protrusions when an adult protrudes his tongue but do not produce more mouth openings when an adult opens his mouth. How can one explain the negative mouth opening results as due to insensitive experimental methods when the same methods in the same experiments were sufficient to demonstrate positive results with tongue protrusion? And why would this pattern of results—tongue protrusion replicated and second behavior not replicated—be common to so many experiments by so many different investigators?

There is something special about tongue protrusions. What is special is that babies stick out their tongues a lot. As a consequence, numerous developmental psychologists have increased their tendency to stick out their tongues at babies. The question arises: Who is imitating whom? Clearly, however, to argue convincingly that infants are not truly imitating adults, we need to know more about why babies stick out their tongues. Is there a simple explanation for the available data that does not rely on imitation?

Answering this last question requires only that we *consider the possibility* that the imitation story is false. For many with an investment in that story, however, even entertaining this possibility is no longer an option. That said, science progresses most assuredly when scientists remain skeptical of their own ideas and systematically rule out alternative explanations until only one reasonable explanation remains. In today's high-speed race for fame, such skepticism is the scientific equivalent of driving in the slow lane.

Returning, then, to imitation and focusing on tongue protrusion as providing the clearest example of an imitated behavior, let's take a few steps back and ask: How often, under what circumstances, and why do infants stick out their tongues? These are very basic questions that were not addressed until the mid-1990s when developmental psychologist Susan Jones tested several hypotheses that follow from a logical progression of simple ideas. First, building on the notion that infants stick our their tongue when they become interested and aroused, she hypothesized that *anything* that stimulates interest and arousal should be followed by tongue protrusion. Second, it follows that infants might appear to imitate tongue protrusions if, coincidentally, tongue protrusions are more interesting and arousing to infants than mouth openings or finger movements. Finally, Jones wondered whether tongue protrusions are so common among infants because the tongue is so central to the infant's early sensory exploration of the world; if so, then it would follow that imitation of tongue protrusion should disappear as other features of exploratory behavior, such as reaching for objects, begin to emerge.

Jones tested her first hypothesis by presenting infants with blinking lights, using a railway signal from a toy train set. Although blinking lights bear little resemblance to a protruding tongue, infants were much more likely to stick out their tongues while looking at the blinking lights than when looking away. Moreover, they stuck out their tongues at the blinking lights more often than they opened their mouths. Thus, infants do stick out their tongues more often when aroused by an inanimate visual object. Moreover, visual stimuli are not necessary; for example, Jones recently observed that the overture from *The Barber of Seville* also does the trick.

Jones next examined why infants stick out their tongues more often when an adult sticks out his or her tongue, but less often when an adult merely opens his or her mouth? Consistent with her hypothesis, Jones found that infants look longer at and, thus, appear more interested in, faces with a protruding tongue than they do at faces with an open mouth. In other words, what looks like imitation from one perspective looks a lot like a coincidence from another.

Finally, Jones's approach is powerful enough to address a thorny issue for the imitation enthusiasts: infants cease imitating tongue protrusions within a few months after birth. If imitation is "innately wired," as Meltzoff would say, then why would it disappear even as it lays the foundation for such later-developing capacities as empathy and mind reading? Jones answered this question by examining the behavior of Nathan and Justin, two infants who were tested repeatedly between the ages of three and thirty weeks. Each week, a toy was placed in front of the infants and their behavioral responses were videotaped. Nathan did not use his arms to reach for an object until he was twelve weeks old. Before this milestone was reached, he protruded his tongue when the object was presented; once he began reaching, however, tongue protrusions became rare. In contrast to Nathan, Justin did not exhibit his first reach for a toy until he was twenty weeks old; nonetheless, like Nathan, he ceased protruding his tongue as he developed his capacity to reach. Jones interprets this parallel between the two children as suggesting that infants protrude their tongues initially because they "cannot execute the full sequence of behaviors that will bring the object to the mouth," but that the onset of reaching "puts the final component of autonomous oral exploration into place" and brings with it the ability "to grasp and transport objects to their mouths, where tongue movements can fulfill their true exploratory function."

One might think that Jones's rigorous and clear-headed approach and Anisfeld's incisive critique would force a reconsideration of infant imitation. But to think so would be a mistake. In fact, their work has been virtually ignored by Meltzoff and by the wider community of developmental psychologists. There are many reasons for this, including the allure of the story itself. But this is not the first alluring story that has turned out to be wrong, and it won't be the last.

## *Running Numbers*

We can learn only so much about the infant mind by way of the tongue. So, to gain further access to the infant mind, developmental psychologists have taken advantage of the infant's ability to look at objects. But what can be inferred from an infant that looks more at one object than at another? That is the crux of the issue that I will now explore.

Amazingly, virtually every claim by nativists regarding the remarkable, unexpected competencies and core knowledge of infants is based on experiments using the eyes as portals to the mind; yet, few people are aware of how the resurgence of nativism rests on little more than a bold decision to play fast and loose with an experimental procedure that has been used since the 1960s to answer legitimate questions about infant perception. To appreciate what went wrong, it is important to review what this looking procedure is and what it is not.

Imagine that we want to determine whether three-month-olds, who cannot talk, can distinguish between two colors, say red and green. We sit the infant down in front of a blank television screen. The screen turns red for several seconds, during which time the infant looks at it. Over the next few presentations of the red screen, we may find that the infant looks even longer than it did on the first trial, reflecting what is called a *familiarity preference.* With further repeated presentations of the red screen, however, we notice that the infant looks at the red screen less and less with each presentation. When the infant's looking time has stabilized at a low level, we say that the infant has habituated to the red screen. At this point, we present a green screen on the television and find that the infant now looks at this new color for a long time, similar to its looking response to the first presentation of the red screen. We say that the infant has dishabituated to the green screen and thereby expresses a *novelty preference.*

What can we conclude from such an experiment? Assuming that we have been careful to ensure that the stimuli presented to the infant differed only on the dimension of color (and not, for example, brightness), we can safely conclude that three-month-olds are able to discriminate red from green. This conclusion is firmly anchored by our understanding of how the visual system works. We know, for example, that color vision begins with the stimulation of color-sensitive neurons in the retina and ends with color-processing systems in the brain. Using this habituation-dishabituation paradigm, psychophysicists have explored many aspects of visual system function in infants, including the infant's ability to discriminate motion, line orientation, and three dimensions. The best examples of these kinds of experiments are characterized by exquisite attention to the visual stimuli that are presented to the infants so as to ensure that all variables known to influence looking behavior are carefully controlled.

Having completed the red-green discrimination experiment, we might wish to go further and suggest that the infant, previously habituated to the red stimulus, was surprised upon presentation of the green stimulus. We might suggest that the infant's expectations were violated when the green stimulus was presented. We may even suggest that the infant knows what color is, that she believes that color exists, and that she possesses a color concept. Finally, noting that three-month-olds are closer to birth than to death, we might even declare that possession of a color concept is innate. In fact, there is nothing to stop us from taking these interpretational leaps if we discard the notion that science should always favor the simplest sufficient explanation (parsimony), and if we decide that we do not care that our explanation cannot be proven false (falsifiability). Freed from such constraints, we can concoct any explanation we please for the infant's ability to discriminate color.

Nativists do not concern themselves with such "mundane" issues as color discrimination. They have much bigger fish to fry. The basic question is whether their pursuit of the bigger fish justifiably takes them out of the realm of the perceptual and into the realm of the conceptual. They think it does. The evidence says otherwise.

We can consider first the question, beloved by nativists, that our understanding of number and numerical computation rests on an innate core capacity that can be detected early in infancy. One promoter of this idea, Karen Wynn, proposes that there

> exists a mental mechanism, dedicated to representing reasoning about number, that comprises part of the inherent structure of the human mind. A range of warmblood [sic] vertebrate species, both avian and mammalian, have been found to exhibit numerical discrimination and reasoning abilities similar to those documented in human infants. Because of its adaptive function, this mechanism quite likely evolved through natural selection, either at a point in evolutionary history prior to the branching off of these different species, or separately but analogously within several branches of the species.

This is quite an impressive collection of undemonstrated, premature, and untestable ideas, arising from just a handful of habituation-dishabituation experiments in human infants. To inspire such grand theorizing, one might expect that these experiments—which Wynn believes demonstrate that five-month-olds "appreciate the precise numerical relationships that hold between small numbers"—are unassailable.

In one seminal experiment conducted by Prentice Starkey and Robert Cooper, five-month-old infants were repeatedly shown stimuli containing two dots until they habituated, that is, reduced their looking time below some criterion. After habituation, when they were shown a stimulus containing three dots, the infants dishabituated, that is, they increased their looking time to this new stimulus (other infants were habituated with three dots and dishabituated with two dots). Because these investigators apparently ruled out such possible factors as the spacing of the dots and their brightness, Wynn praises this experiment for its demonstration that infants "possess a sensitivity to number *per se*," that is, that their sensitivity cannot be explained by any account that appeals to "low-level perceptual mechanisms," such as dot spacing, brightness, and the like.

The notion that low-level perceptual mechanisms are inadequate to account for the infants' performance on these tasks is central to the claims that nativists wish to make about the innate competencies of infants. Nativists do not want mere perceptual mechanisms to be at play because they have their have set their sights on a much bigger trophy than mere perception. They are trying to bag core knowledge. Perception, apparently, is for little people.

Unfortunately, the nativists' haste to claim their prize can blind them to critical flaws in their experimental procedures. For example, developmental psychologists Melissa Clearfield and Kelly Mix wondered whether one important variable was being overlooked. That variable is contour length, which is the length of the perimeter around an individual object. For example, a square with one-centimeter sides has a contour length of 1 x 4 sides = 4 centimeters. Because research dating back to the 1960s had shown that infants are sensitive to contour length, it seemed only reasonable that this was a factor that needed to be ruled out before grander conclusions should be entertained.

What Clearfield and Mix did was repeat the habituation-dishabituation experiments of earlier investigators but, this time, they examined whether infants would dishabituate to a stimulus in which contour length was changed but not number, and whether infants would dishabituate to a stimulus in which number was changed but not contour length. If contour length is the key variable, then infants should only dishabituate to contour length even when the number of items changed. This is exactly what Clearfield and Mix found, demonstrating that "infants attend to contour length, rather than number, to discriminate between sets." But, in contrast with the nativists whose work they were correcting, they sounded a note of caution by stating that "because contour length is correlated with total area, brightness, and size, it could be any or all of these variables that affect infants' looking behavior in this task." At the very least, they concluded, "infants prefer to discriminate on the basis of basic perceptual variables, rather than relying on abstract cognitive knowledge."

Contributions such as that of Clearfield and Mix are undervalued by the scientific community. Whereas the original Starkey and Cooper report on the "perception of numbers" was published in the premier journal *Science,* Clearfield and Mix's contour length study was published in a prominent but much less prestigious psychology journal. This is an unfortunate but common occurrence, especially with regard to extraordinary scientific claims: the big splash is made in the prestige journals, attendant with extensive media coverage, and the mess must be cleaned up later by more careful researchers after the cameras have been turned off and the incorrect information has already disseminated into the public's mind and filtered into textbooks. Indeed, when Steven Pinker inventoried human instincts ten years ago, an understanding of number was listed third.

When Clearfield and Mix published their findings, there were no banner headlines proclaiming that INFANTS POSSESS INNATE KNOWL-EDGE OF CONTOUR LENGTH. Why? First, as already mentioned, the idea that infants are sensitive to contour length was not new. Second, researchers who investigate such things as sensitivity to contour length seem less attracted to the allure of nativism. And third, there is no prestige to making nativist claims about low-level perceptual mechanisms; such cachet is reserved for those skills that provide evidence of infant conceptual understanding. But as we have seen with number concepts, conclusions pertaining to infant conceptual knowledge rest on shaky ground. In the end, there is no substitute for rigorous experimentation and thoughtful interpretation.

The nativist interest in number has been taken even further in a series of experiments aimed at demonstrating the ability of infants to add and subtract. Given that numerical computation rests on an understanding of number, and given the recent unraveling of the claim that infants possess an understanding of number, there is little need to delve too deeply into these experiments. Nonetheless, some discussion here is warranted.

The basic approach, made famous by Karen Wynn in a paper published in 1992 in *Nature*, is to seat infants in front of a small doll-sized stage. From one wing of the stage, a hand enters and a Mickey Mouse doll is placed there (the addition of the doll to the empty stage represents the calculation 0 + 1). The hand disappears and a screen, which blocks the infant's view of the doll, is lifted. Then, to the side of the screen and in view of the infant, the hand reappears with another doll, the hand and doll disappear behind the screen, and the hand leaves the stage without the doll (1 + 1). At this point the screen drops to reveal one of two scenes: First, in the "possible" condition, the infant sees two dolls, as would be expected; this condition represents the calculation 1 + 1 = 2. Second, in the "impossible" condition, the infant sees only one doll, which represents the calculation 1 + 1 = 1; obviously, an infant that understands numerical calculation would be surprised by such a scenario. Other calculations can be performed using this paradigm, including the calculation 2 - 1 = 1. What Wynn reported was that infants looked longer at the "impossible" or unexpected scenes because these scenes violated the infants' innate knowledge of the rules of numerical computation. Accordingly, infants "appreciate the precise numerical relationships that hold between small numbers" and it is this appreciation that provides infants with a "toe-hold upon which to enter the realm of mathematical thought."

These are heady conclusions indeed, but are they warranted? If we step back and reconsider the story that culminated in Clearfield and Mix's critique of infants' sensitivity to number, we should appreciate immediately the immense complexity of Wynn's doll experiment. Instead of dots on a page, we now have a stage and dolls and a screen and hands appearing and disappearing. By the time we have arrived at the end of the test and the screen drops to reveal to the infant the final critical scene, it is clear that we are dealing with a level of complexity that dwarfs the experiments described above. For Wynn, this complexity provides the necessary richness to reveal the high-level cognitive capacities of infants. But for those who seek to replicate Wynn's experiments and critically analyze her claims, this complexity muddies the waters and thwarts attempts at clarification.

Nonetheless, Wynn's doll computation experiments fail to convince for a number of reasons. A very brief survey will suffice here. First, whereas some of Wynn's experiments have been replicated, others have not, and the pattern of replications tells an interesting story. The most robust of Wynn's findings entails experiments in which the "impossible" or incorrect answer is associated with the final scene containing more dolls. This is a serious concern because other investigators have shown using similar methods that, even in the absence of numerical calculation, infants prefer to look at scenes that contain more dolls. In addition, there is the simple matter that adding and subtracting the number of dolls in a scene entails changes in surface area, volume, and, yes, contour length. Even an experiment by fellow nativist Elizabeth Spelke and her colleagues bears out this last concern: when the size of the dolls is controlled in a numerical computation experiment similar to Wynn's, the infants display a sensitivity to size, not number.

Wynn's procedure is typical of so many that have been used by nativists to argue for the "amazing hidden competencies" of infants. What distinguishes these experiments from those of the psychophysicists, who focus on basic perceptual capacities, is the marked increase in the complexity of the experiments. Thus, Spelke examines infants' understanding of the solidity of objects by dropping balls behind screens and determining whether infants look more at those scenes in which the ball appears to have passed through a solid floor. Renée Baillargeon also performs a test of solidity but uses a "drawbridge" that appears to pass over and through a solid object that, in the real world, would obstruct its path. Similar experiments have been performed to test infants' understanding of causality, gravity, and support. But although the details may change from experiment to experiment, there is a common thread that joins them all. This thread, simply put, is the tendency of infants to look at visual scenes and the tendency, shared by all animals, to habituate and dishabituate to stimuli. It really is not more complicated than that.

The late developmental psychologist Esther Thelen was a leading critic of these nativist experiments and, working with Gregor Schöner, sought to place infant behavior in these tasks on firmer scientific ground. They created a "dumb" mathematical model that captures the looking behavior of infants while discarding the fanciful nativist and mentalistic interpretations. The advantage of thinking mathematically about a problem such as this is that it forces consideration of every feature of an experiment that can influence the infant's looking behavior at the end of the test. In this light, the nativist approach is found to be lacking on the critical dimension of experimental rigor. As Schöner and Thelen observe,

> seemingly very small changes in stimuli, in the timing of presentations, in the metric differences of the test events, in the number of trials and habituation criteria, and in individual infants can radically change the outcome measure of whether a baby dishabituates to a test stimulus. In many (or even most) infant habituation studies these factors are not well specified, and often are unknown and unreported.

This is a damning statement when one considers the extraordinary claims that nativists, flying in the face of common sense, have made on behalf of their mute experimental subjects. The skill of nativists has been to borrow a valuable experimental method for examining the perceptual abilities of infants, increase the level of complexity, decrease the level of experimental control, and engage in unconstrained cognitive and evolutionary theorizing. The result may seem impressive to some, but it is a house of cards. As Schöner and Thelen observe, researchers "may invent experiments to create 'violations of expectancy' but infants experience them as visual events." In other words, nativists are leading their witnesses.

## *A Reach*

The nativist stance has been uncritically accepted by so many psychologists because it fits seamlessly with other intellectual trends within the field. One such trend can be traced to the unleashing of cognitivist approaches to psychology in the 1960s as the behaviorist tradition lost its grip and interest in behavior was displaced by interest in the mind. The nativist twist was to substitute an individual's accumulated knowledge with a species' evolved knowledge, a twist that gives license to imagine that infants are born with a competence that they cannot express. Once this distinction between competence and performance was accepted, the task of the nativist was clear: devise experiments that help the infant to overcome its inability to perform.

Of course, in infants as well as in adults suffering from degenerative neurological diseases, *some* form of behavior is required for the mind to communicate with the outside world (although computer-assisted prosthetics may one day permit translation of brain activity directly into behavior). For adults in need of help, highly sensitive devices allow computers to translate finger or eye movements into language. For infants, refined methods—such as the looking paradigm described above—are needed to reveal infants' hidden competencies. There is, however, a crucial difference between the two circumstances. When an adult uses a computer-assisted device, the communicative act is direct and transparent;

but when an infant "communicates" with its eyes in a looking task, we must rely on the experimenter to translate the infants' behavior accurately. As we have seen, much can be lost—and added—in translation.

We have, then, two competing perspectives of the relation between cognition and behavior as they apply to infants. One camp sees a dichotomy in which the mind and body stand apart. Accordingly, it is argued that there exists a central unchanging core of knowledge around which the behavioral competence of the infant grows. Another camp views this dualistic dichotomy as false, arguing for the notion that cognition is embodied—that mind and body are inseparable—and, therefore, any experiment designed to probe the mind must pay close attention to the behavioral task that infants are being asked to perform.

There are several consequences of the competence-performance distinction that concern us here. First, it cannot be overemphasized that the nativist stance rests on the assumption that a fragment of a competence can be substituted for a full-blown competence. For example, we all know that infants will kick their legs in an alternating fashion when placed on their backs. Esther Thelen performed groundbreaking work on the development of locomotion by investigating the infant's alternating leg movements and their contribution to upright walking. Her work was distinctly non-nativist in that it sought to piece together the various components that make walking possible.

In a crafty twist, Elizabeth Spelke and Elissa Newport argue that Thelen's work actually supports the nativist perspective. To make their case, they first draw attention to the fact that alternating leg movements occur before infants are able to maintain an upright posture and walk. Then, they note that this pattern is "present before it is needed for that purpose." Finally, they refer to the alternating leg movements as a "competence that does not normally express itself in behavior, but that can be elicited under appropriate supporting conditions." Note the bait and switch, the easy glide in which a basic movement *pattern* is transformed into an early *competence*. This is a distortion, common to many nativist arguments, that rests on two fundamental errors: First, there is the confusion of the fragments of a competence with the full-blown competence itself. It should be obvious that moving your legs is not walking just as moving your eyes is not reading, moving your fingers is not typing, swinging a bat is not batting, moving your hips is not dancing, and scratching your head is not thinking.

The second error derives from an unwillingness to appreciate just how idiosyncratic are the methods used to reveal the infant mind. When Spelke and Newport refer to "appropriate supporting conditions," they are suggesting that younger infants require an external means of support—a scaffold—to allow them to express their hidden, inborn capacities. The younger the infant, the greater the need for support. The error comes in assuming that the scaffold aids the infant's performance but has no effect on its competence.

To appreciate the significance of this last point, imagine that you are standing on a balance beam. As you position yourself on the narrow wooden beam, you become acutely aware of those positions of the feet, hips, and arms that achieve the most secure position. When your hips move in one direction, you compensate with movements of your upper body in the opposite direction. But now imagine that four people are supporting you from below, bracing your feet, legs, hips, and arms; if these helpers move your body in one direction, you are free to compensate by moving your head in the opposite direction. Such compensatory head movements may indicate the presence of one component of a multicomponent system that helps you to achieve balance, but we would never confuse *performance* of this single component of balance-keeping with the *competence* to walk on the beam without support. Thus, the two errors committed by nativists—mistaking the parts for the whole and discounting the importance of how infants are tested—combine to create the illusion of early competence where, at best, only the fragments of competence actually exist.

So far we have examined two simple behaviors—tongue protrusion and eye movements—on which developmental psychologists have relied to explore early infant behavior. Despite the nativist desire to layer ornate interpretations on top of these simple behaviors, we have seen that more careful analysis of the behavior undermines these interpretations. To bring this point home, consider a recent controversy that surrounds a simple hide-and-seek game first described by Jean Piaget in the 1950s.

Piaget viewed development as a series of stages through which infants pass as they develop. He identified the fourth stage as spanning the period of approximately seven to twelve months, when infants fail to appreciate that objects continue to exist even when they are no longer in sight. To demonstrate this failure of object permanence, he described a game in which an infant sees Piaget hide his watch beneath a cloth to the infant's right. The infant lifts the cloth and recovers the watch. This cycle of hiding (always to the right of the infant) and retrieving is repeated a few times, and then Piaget changes the game by hiding the watch under a second piece of cloth to the infant's left. Despite having watched Piaget make the switch, the infant erroneously lifts the cloth to the right, that is, the cloth that had previously hidden the watch. This is a very funny error to observe, and Piaget concluded that it demonstrates the infant's inability to understand how objects operate in time and space. The age limit of seven to twelve months reflects the fact that infants younger than seven months do not reach for objects and infants older than twelve months no longer make this particular error.

Over the subsequent decades, Piaget's informal game evolved into a classic experimental task. The mature version of the task has been described like this:

[A]n infant sits before two hiding locations that are *highly similar* and separated by a small distance (e.g., two identical cloth covers or two identical lids that are 20 cm apart). While the infant watches, a desired object is hidden in one location, the A location. After a *delay,* the infant is allowed to *reach* and search for the object. This hiding and search at the A location is *repeated several times.* Then, again while the infant *watches,* the object is hidden at B. After the delay, the infant is allowed to reach. In this canonical form of the task, eight-to-ten-month-old infants reliably reach back to location A, making the A-not-B error.

In this description, the italicized words indicate all those features of the task that are now known to influence the likelihood that an infant will commit the A-not-B error. For example, the degree of similarity of the hiding locations, the delay between the hiding and the reaching, and the number of repetitions to the A location all influence the error. Thus, we know a great deal about those features of the task that influence the likelihood of committing the error.

Still, what does the error mean? One favored interpretation is that commission of the A-not-B error indicates that an infant does not yet understand that objects do not simply appear as a consequence of our looking at them; in other words, error-prone infants have not yet developed a mature object concept. But this is only one of several interpretations that have been considered and tested over the years. Indeed, it is not an exaggeration to say that Piaget's A-not-B task has become an industry all its own, replete with favored methods and interpretations.

Most recently, however, Esther Thelen, Linda Smith, and their colleagues and students focused attention on the task itself and what it demands of the infant perceptually and behaviorally. Accordingly, they asked, "If the A-not-B error is a true measure of the status of infants' representations of objects, how can it be that what they know depends on so many seemingly irrelevant factors? How can it be that infants have a more mature object concept at home than in the laboratory or when the object is a cookie rather than a small toy?" When all of these various anomalous findings are collected and analyzed, they argue, it becomes clear that the A-not-B error "is not about what infants *have* and *don't have* as enduring concepts, traits, or deficits, but what they *are doing* and *have done*." In short, the A-not-B error emerges from *all* the specific requirements of the task—looking, reaching, and remembering.

For example, consider the fact that the A-not-B error only occurs when infants must wait several seconds before they are allowed to reach for the hidden object. If there is no delay and infants are allowed to reach immediately, then the error is not made at any age. Does this mean that an infant possesses an object concept when the object is first hidden, but loses the object concept as time passes before the reach?

Next, consider the role that the infant's prior reaches to A plays in the production of the error. Contrary to the impression given by Piaget's informal account of the error, infants do not readily reach for objects hidden from view. On the contrary, infants must be trained to reach and this training contributes importantly to their developing preference for the A location. In other words, by training infants to move their arms toward the A location time after time, a motor habit is established that is not easily broken when the object is hidden at the B location. Thus, infants return to the A location because they have reached there in the past, not because they lack an object concept.

This last point may leave you incredulous: How could so many developmental psychologists working over so many decades mistake a motor habit for a deficiency in the understanding of objects? Actually, the status of the conventional interpretation of the A-not-B error is even more dire. Consider this astounding fact: infants will commit the A-not-B error even when the experimenters merely wave the lids in front of the infants but *do not hide the toy!* To repeat: a hidden toy is not a prerequisite for demonstrating the A-not-B error.

What would inspire investigators to strip down a classic experimental task so that the universally assumed essential ingredient—a hidden toy—is removed altogether? The answer is that once the focus moves from hidden mental capacities to actual behaviors in an experimental task, novel (and even radical) ideas present themselves. Deleting the toy from the A-not-B task was a radical move but, once performed, it necessitates a reorientation of the history and significance of the task: Piaget's hide-and-seek game loses its conceptual link to object permanence when the object itself disappears.

The deconstruction of the A-not-B task has continued through a number of other ingenious experiments. For example, if the infant's behavior is indeed fragile and linked critically to its prior experience with reaching, then a manipulation as simple as changing the infant's posture should eliminate the error. This manipulation was accomplished by training infants while sitting and then testing them while standing. Needless to say, this change in posture was sufficient to erase the error.

Finally, if the A-not-B error is generated by the specific demands of the experimental task, then it should be possible to revive the error in older children by increasing the difficulty of the task. This was accomplished in children at two and three years of age by, for example, hiding objects in a sandbox (in which exact locations are not as easy to remember) and increasing the delay between hiding and reaching to ten seconds (which places a heavier demand on a child's memory). Under these conditions, children committed the error by reaching back to the A location when the object was hidden at the B location. Thus, Piaget's notion of a discrete stage during which children commit the A-not-B error gives way to a notion of proneness to error that reflects the specific features of the experimental task and the capabilities of the subject.

We have seen how our understanding of infant cognition and behavior is informed by directing the spotlight toward the experimental methods being used. Thus, newborns appear capable of imitating tongue protrusions until we realize that they stick out their tongues whenever they are aroused. Five-month-olds appear to possess a "number concept" until we realize that they pay attention to contour length, not number. And an "object concept" seems to emerge after eleven months of age until we realize that the method used to detect this object concept need not employ objects.

Given the problems that we have seen with the A-not-B paradigm, one might wonder whether looking tasks provide privileged access to the mind that reaching tasks cannot provide. But why would this be the case? After all,

> looking at or away from an event display is a motor act integrating attention, vision, and memory processes just as much as reaching toward one place or another. . . . By this reasoning, neither looking nor reaching provides a direct readout of the contents of the mind. Both are constituted on-line, within the moment, the bounds of the task, and the child's history in similar situations. Looking tasks, like the A-not-B, can tell us a lot about how these processes work together, but they cannot claim privileged access to the enduring contents of mind.

No matter how much we may wish to understand the contents of the infant mind, we cannot divorce any knowledge we might gain from the actual experimental methods that we have at our disposal. Whether we focus on the tongue, the eyes, or the hand, we should not forget that these are anatomical parts whose movements depend upon muscle activity. At best, inferences about the mind are indirect; at worst, they are based on flawed experiments. For many, such problems are mere details that cannot shake the power of the easy sound bite that keeps nativism alive in the public imagination. One such sound bite is this: infants possess innate core knowledge about the world— about continuous motion, solidity, and number—that remains unchanged during development. It is now time to turn our attention to this central nativist claim.

### *You Can't Take It With You*

If it appears that I have disdain for the nativist perspective, it is because I am convinced that it is an intellectual and experimental red herring. Some psychologists—even those who are doubtful of nativist claims—believe that nativists have served science well by challenging empiricists to consider the biological determinants and evolutionary antecedents of knowledge. This position is more generous than true. There is little doubt that developmental psychologists would benefit from a broader understanding of animal behavior, developmental biology, and evolutionary concepts; nativists, however, seem ill-equipped to provide the instruction. Indeed, the conceptual foundation of nativism could not be less biological.

As Elizabeth Spelke states it, there are two general claims that nativists wish to make about human development. The first claim is that infants "are capable of reasoning" and that they "come to know about states of the world that they never perceived." The rationalist underpinnings of this perspective are revealed by the telltale buzz words that pepper the nativist literature. Nativists frequently write that infants *believe, realize, expect,* show *surprise, appreciate, recognize,* and, of course, *reason.* It is important to stress that these words—for which no adequate definitions are given—are meant to explain infant behavior, not merely describe it. One nativist, Renée Baillargeon, is particularly clear on this point:

> I challenge anyone to explain [my] findings without ascribing to infants "high level concepts" such as a belief (whether innate or not) in the continued existence of occluded objects, and rather sophisticated reasoning or inferential abilities; it simply cannot be done.

This is a bold assertion that deserves a bold retort: whenever one encounters an explanation of infant behavior that relies on high-level concepts such as beliefs and expectations, it is safe to assume that it is wrong. There is now ample evidence—some of it already discussed in this chapter—to support this assertion.

Spelke describes a second general nativist claim, namely, that "development leads to the enrichment of conceptions around an *unchanging core*." This contention of an essential, irreducible quality of conceptual knowledge is yet another manifestation of the marriage of nativism with rationalism. It is a powerful and deeply engrained idea that is not likely to go away soon. But there are many reasons to doubt its validity. Here, I consider two situations where knowledge fails to transfer from one experimental situation to another. As will be shown, knowledge is not nearly as portable as nativists contend.

Newborns have no choice but to quickly adapt to the physical world and the many forces that act within it. Using the looking task described above, nativists have explored the infant's ability to detect physically

through solid floors. These experiments purportedly demonstrate that infants "represent physical objects" and "reason about object motion" but, for reasons already discussed, these experiments have numerous flaws and alternative explanations. Nonetheless, many consider these nativist claims as fact. For example, on his list of human instincts, Steven Pinker's first entry is, "Intuitive mechanics: knowledge of the motions, forces, and deformations that objects undergo."

If infants instinctively understand and reason about the physical forces acting on objects, and if these instinctive capabilities contribute to survival, then wouldn't we expect infants to avoid situations that are physically threatening? After all, what is the point of having an intuitive understanding of physics if this understanding does not help you to avoid maiming or killing yourself in a fall?

The body of a human newborn is loaded on top with an oversized head and neck muscles that are unable to support it. Each day, body parts grow at different rates and muscles strengthen. Meanwhile, the infant experiences the world from dramatically different perspectives as, by six months of age, sitting is possible; by eight months of age, crawling begins; by ten months of age, infants can stand up and walk using furniture for support; and by twelve months of age, they are independent walkers. This developmental trajectory of human locomotion illustrates why so many developmentalists question the validity of the nativist perspective: How could an unchanging core of physical knowledge possibly prepare the infant for the rapidly changing worldly experiences that accompany these dramatic transitions from stationary to ambulatory, from sitting to standing, and from dependence to independence?

If infants possess an intuitive knowledge of physics, it would provide no better service than to inform the infant's decisions regarding safe and unsafe locations and situations. The infant's ability to avoid dangerous situations was first examined many years ago using a "visual cliff" (an apparatus comprising a glass platform spanning an abyss such that an infant can see the drop-off without being placed in danger). Interestingly, avoidance of the cliff does not begin until after infants have had substantial experience with crawling—after seven months of age—despite the fact that much younger infants are able to perceive depth. Thus, the visual cliff seems to indicate a dramatic failure of the thesis of core knowledge: experience with crawling, not an innate fear module, appears critical for cliff avoidance. In response, of course, the committed nativist will counter that a fear of heights is innately timed to coincide with the burgeoning ability of infants to expose themselves to danger.

Granting this last point for the sake of argument, we would nonetheless expect a nativist to argue that avoidance of dangerous situations, once demonstrated, would reflect an unchanging core of the infant's knowledge about the physical world. Indeed, a nativist would likely argue further that, once this knowledge is acquired—innately or otherwise—an infant should be able to use its reasoning abilities to assess similarly dangerous situations in the future. In fact, however, infants are surprisingly poor at reasoning about their knowledge of danger in the physical world.

Moving beyond the visual cliff to more realistic challenges, developmental psychologist Karen Adolph has studied the behavior of infants walking up and down dangerous slopes. Using an apparatus constructed of two flat landing platforms and an adjustable incline, Adolph placed infants on the upper platform and encouraged them to descend. By testing the same infants repeatedly every three weeks across the transition from crawling to walking (from five to sixteen months of age), Adolph assessed the infants' developing locomotor skills as they gained increasing experience with different modes of locomotion.

Not surprisingly, as infants grew and as their level of skill developed, they were increasingly able to descend steeper slopes without falling down or without having to be rescued by the experimenter. Infants also became better judges of their skill level, but only after many weeks of experience with their current mode of locomotion. Prior to this, however, infants would fearlessly and recklessly plunge over slopes that were impossible for them to descend successfully. The surprise came when Adolph compared the behavior of infants at ages when walking was just beginning to replace crawling as the dominant mode of locomotion:

> [I]n the first weeks of walking, the same babies attempted to walk down the same impossibly risky slopes that they had so recently avoided in the crawling posture. In fact, new walkers showed no transfer from their old, familiar crawling posture to their new, upright walking posture on consecutive trials at the same risky slope. Over weeks of walking experience, errors decreased but learning was no faster the second time around.

This is remarkable, but familiar. You may recall that infants perform differently on the A-not-B task when they are trained in a sitting position but later tested while standing up. But why should this be? Adolph explains that learning "may be posture-specific because each postural milestone represents a different perception-action system with different control variables." When sitting, crawling, and walking, infants are placing primary stress on different joints—hips, wrists, and ankles—and relying on different muscle groups. In addition, infants in these postures experience the world from "different vantage points for viewing the ground; different patterns of optic flow as the body sways back and forth; different correlations between visual, kinesthetic, and vestibular information; and so on."

Adolph recently revisited the visual cliff experiment but altered the task so that infants, while in a sitting or crawling posture at the edge of a cliff, were encouraged to reach across an adjustable gap. Using nine-month-olds who had greater experience with sitting than with crawling, Adolph examined the willingness of the infants to reach across wide gaps and risk falling into the crevice. As with the slope experiment, infants' judgments were posture-specific: Infants who accurately judged a gap as being too large in a sitting position showed no reticence when tested that same day in the less-familiar crawling position.

This last finding is very important because it belies the notion that knowledge about physical danger is like a universal form of currency that purchases safety when needed. Rather, knowledge appears to be highly specific to past experiences and present circumstances. Clearly, infants do not possess some generic form of knowledge. As Adolph notes, "If infants learn to avoid a drop-off because they are afraid of heights, know that cliffs are dangerous, or know that their bodies cannot be supported in empty space, then they should show similar adaptive avoidance responses to a drop-off regardless of the posture in which they are tested." And what are we to make of the nativist claim that infants possess an intuitive knowledge of physics and gravity when this knowledge is not sufficient to prevent an infant from walking over a cliff?

So how do infants learn to make accurate judgments about physical risk? In neither the slopes nor the gaps experiments was there any evidence that infants learn much from falling down, whether in the home or in the laboratory. Instead, Adolph argues that

> learning about balance control may occur over many thousands of trials during the course of each day. . . . To put the sheer magnitude of infants' locomotor experience into perspective, each day crawling infants practice keeping balance for more than five hours; during that time, they can travel the lengths of two football fields and take more than three thousand crawling steps. Walking infants practice keeping balance for more than six hours per day; during that time, they can travel the lengths of twenty-nine football fields and take nine thousand walking steps. In addition to the time that they are in transit, crawlers, cruisers, and walkers gain hours of daily experience maintaining balance in stationary positions. Each little crawling and walking step, each forward lean and reach, each time infants roll over or sit up is a tiny "trial" with balance control.

Nativists may find all of this to be terribly inefficient. They might insist that it would be better for infants to possess innate core knowledge about those situations that pose real, significant threats to our health and safety. Indeed, on Pinker's list of human instincts, the sixth entry is "danger, including the emotions of fear and caution" and "phobias for stimuli such as heights." There is, however, a clear and compelling reason why such notions are unrealistic and, ultimately, inefficient. Simply put, physical danger is not an absolute. What is dangerous one day is safe the next as bodies, muscles, and nervous systems change, and as infants' increasing mobility exposes them to new and unpredictable environmental challenges; any programmed "danger rules" that might work one moment would be obsolete the next.

Nativists may cling to the notion that infants begin their lives with core knowledge that is elaborated during development, but such a notion cannot account for the poor judgments of infants at the transition to each new postural milestone. What appears, then, to be most efficient is the continuous calibration of behavior to the needs of the moment based upon experience accumulated through the mundane activities of everyday life.

The nativist belief in an unchanging core of knowledge subsumes an even deeper assumption, namely, that infant knowledge transcends the here and now—that knowledge is portable, static, and certain. On the contrary, Adolph's work clearly demonstrates that, contrary to the assumptions of nativists, even knowledge that has obvious survival value is experience-dependent and context-sensitive.

Another avenue for examining the portability of knowledge across development is to assess the ability of infants to use language to discuss a memorable event that occurred before the development of verbal skills. If knowledge is a discrete entity that infants possess, then that knowledge should be easily transportable from a preverbal to a verbal context. On the other hand, if knowledge is sensitive to the conditions prevailing at the time of encoding, then nonverbal memories should be impervious to the subsequent enhancement of verbal skills.

This issue was addressed experimentally by playing a unique game with preverbal and verbal children and then testing their memories for the game six or twelve months later. The preverbal children were twenty-seven months old and were shown, through testing, to have limited verbal skills and, most importantly, to know very few words associated with the game that they played. The verbal children were thirty-three and thirty-nine months old and were shown to have advanced language skills compared with the twenty-seven-month-olds. In addition, the preverbal children's language skills increased significantly when tested six or twelve months later.

Initially, experimenters played the game twice with each child on successive days; by the second day, the children were able to play the game independently (that is, they were able to perform all of the correct actions in the correct sequence). When tested six or twelve months later, their nonverbal memories for the game were examined by having them identify objects in photographs associated with the game or having them reenact the game behaviorally. Based on these tests, it was clear that the children at all three ages had strong nonverbal memories both six and twelve months after playing the game.

With regard to verbal memories, the children who experienced the game at thirty-three and thirty-nine months of age were able to answer specific questions both six and twelve months later. The critical test, however, came when testing the children who first played the game at twenty-seven months of age. As Gabrielle Simcock and Harlene Hayne describe their results,

> there was not a single instance in which a child used a word or words to describe the event that had not been part of his or her productive vocabulary at the time of encoding. Thus, although the children could remember information that had been encoded without the benefit of language, they could not translate that information into words even though they had acquired the vocabulary to do so. . . . In short, children's verbal reports of the event were frozen in time, reflecting their verbal skill at the time of encoding, rather than at the time of test.

These results do not relate directly to the types of innate core knowledge that nativists discuss because the knowledge gained by playing a game is clearly learned through experience. On the other hand, nativists wish to convince us that infants can reason about their experiences using the unchanging core knowledge with which they are born; moreover, nativists contend that reasoning about core knowledge forms the foundation for our mature conceptions of how the world works. The Simcock and Hayne experiment, however, demonstrates that even children beyond two years of age cannot reason sufficiently about preverbal knowledge to translate it into verbal knowledge. Therefore, in conjunction with Adolph's locomotor studies, we see that knowledge does not, as nativists contend, accumulate according to rules that we, as language-competent adults, may wish to project onto preverbal infants.

### I've Grown Accustomed to Your Face

With the exception of our foray into imitation in newborns, the vast majority of the work described in this chapter concerns infants beyond the age of three months. Nonetheless, on this basis alone, claims of innate core knowledge are made. Is this, however, a convincing demonstration of innateness? With so many common definitions of *innate*—present at birth, not learned, present before its function is needed, arising according to a fixed developmental plan—it is difficult to know which meaning is being invoked at any given time. But even when nativists claim that experiments using three-month-olds provide evidence for innateness, they argue that the burden falls on others to provide alternative explanations.

This is a clever strategy: perform experiments on infants with over ninety days of postnatal experience and, upon the discovery of a competence, claim that this competence is present at birth. When others object, argue that they have the burden of proving you wrong. But is this a fair assessment of who has the burden of proof? An infant with one hundred days of postnatal life has been awake for three million seconds and has moved its eyes approximately three to six million times. When considered in this light, would it not be extraordinary if these millions of eye movements had no effect on the infant's performance on a task that relies exclusively on looking time?

As we saw with the development of pecking accuracy in laughing gulls, only a few days of posthatching experience produce dramatic improvements in behavior. Such rapid developmental effects are commonly seen by those studying nonhuman animals. Why, then, are nativists convinced that, in humans, ninety days of postnatal experience are irrelevant to the competencies that they study? One answer to this question is that it is difficult to investigate the role of experience during development, in part because ethical considerations preclude experimental manipulation of an infant's rearing environment. Certainly, nativists have benefited from the protection afforded by these realities. Nonetheless, several experiments have successfully examined the role of experience in cognitive development and nativist claims have been empirically tested. I will now examine two such examples.

From the spot in my living room where I am writing these words, I can look through my window at the house across the street. Despite the many objects that impede my view—the window panes, the tree in the front yard with its numerous branches—I do not perceive my neighbor's house as a disjointed mosaic, but rather as a unified whole. In the words of Philip Kellman and Elizabeth Spelke, we are constantly surrounded by objects that are only partly visible to us and yet we "perceive a world of complete and solid objects, not visible fragments." But when do we acquire this ability to conceive of fragmented objects as wholes? Based on their study, Kellman and Spelke claim that this conception of unity and coherence is innate.

The original support for this claim emerged twenty years ago from a clever extension of the standard habituation-dishabituation looking paradigm. Four-month-old infants were habituated to a stimulus in which a rod was seen to be moving behind an opaque, rectangular block; only the parts of the rod—at the top and bottom—that were not occluded by the block could be seen by the infants. After the infants had habituated to the rod moving behind the block, they were tested using two displays, both of which lacked the occluding block. One display was a complete rod which, from our adult perspective, would match the object that was likely moving behind the block; the second display comprised the two sections of the rod that were not occluded by the block and that were, as already mentioned, the only parts of the rod that the infants *actually* saw. The question posed by the experimenters was whether the infants would behave as if they mentally completed the unseen portions of the rod and therefore perceived the rod pieces as part of a single, unified object or, alternatively, would they behave as if they saw the rod as two disconnected pieces moving in synchrony on both sides of the block.

When infants were habituated to the occluded rod, they looked longer during the test trial when presented with the two rod pieces than when presented with the single, completed rod. Because it is thought that habituated infants look longer at *novel* stimuli during a test trial, the above result was interpreted as indicating that infants are able to mentally complete occluded rods. Accordingly, Kellman and Spelke concluded that the infants' tendency to complete objects that share perceptual attributes—such as synchronous movement behind an occluding object in the foreground—"roots perception of partly hidden objects in an *unlearned* conception of the physical world" and that this conception about objects "may be an essential property, at the center of our conception of the world."

Subsequent work, however, most notably by Alan Slater and his colleagues using newborns, failed to garner support for the nativist interpretation of the above experiment. Whereas Kellman and Spelke's results were successfully replicated in four-month-olds, newborns behaved as if they had perceived the rod pieces as pieces, that is, they treated "the visible evidence literally." In other words, these newborns did not fill in the gap behind the block as did the four-month-olds. Clearly, this result contradicts the favored nativist view "that infants from birth have an innate idea of the underlying unity and coherence of objects."

▪ ▪ ▪

The ability to recognize a face is, one might think, much more complex than completing an occluded object, requiring visual processing of specific anatomical features as well as their spatial relations. Face recognition is a fundamental feature of human cognition that is presumed to have important adaptive implications for social behavior. In addition, face recognition is thought to rely on specialized neural processing. For example, damage to a constrained region of cerebral cortex in adults produces a syndrome called prosopagnosia, which is characterized by the selective inability to recognize faces. Thus, it is not surprising that nativists contend that face recognition is an innate capacity that is made possible by a dedicated neural module.

Related to the expert abilities of infants to recognize human faces is their apparent preference for female faces. This was first demonstrated a decade ago using infants between the ages of five and twelve months. More recently, developmental psychologist Paul Quinn and his colleagues demonstrated this bias in infants as young as three months of age. The preference for female over male faces was easily demonstrated. Infants were shown pairs of color photographs of female and male faces simultaneously and the experimenters measured the amount of time that the infants looked at each face in the pair. Infants looked significantly longer at, and therefore presumably preferred, the female faces.

The color photographs used to demonstrate this preference of infants for female faces were cut out from a clothing catalogue. Photographs were selected such that the facial expression of the models, their attractiveness, as well as the orientation of the faces did not influence the results. Nonetheless, the female faces could have attracted the infants' attention more readily because of the greater amounts of hair surrounding them. So, the experiment was repeated with the same facial stimuli but with all hair removed. The results, however, were the same: infants preferred the female faces. Other similarly trivial explanations were also discounted.

How can we explain such a robust and early preference for female faces? It is known that female faces are structurally different from male faces, so perhaps infants have a hardwired sensitivity to these structural differences that allows them to recognize likely caregivers. Because young infants prefer their mother's face to that of a strange female's, the general preference for female faces could even be associated with an innate mechanism for identifying the primary caregiver, that is, mom. Consistent with nativist tradition, we would assume that these preferences are inborn and place the burden on others to prove otherwise.

In this case, proving otherwise is difficult but not impossible. First, we note that infants' preferences for the face of mom or for the faces of females might arise through experience. Simply put, three-month-old infants may exhibit a preference for female faces because those are the faces to which infants are predominantly exposed during their first three postnatal months. Second, if this perceptual experience hypothesis is correct, then it should be possible to create a preference for male faces by exposing infants predominantly to male faces in early development.

This is exactly the point where most nativist claims find safe refuge: we cannot experimentally manipulate the parental exposure of human infants any more than we can rear them in the dark, or in gravity-free environments, or in a world where solid objects do not exist. Luckily, however, there are instances where fathers, not mothers, turn out to be the primary caregivers, and Quinn looked hard to seek out these father-raised infants and test them between three and four months of age. He found eight such infants and, lo and behold, seven of them exhibited a preference for male faces. More infants may be needed to nail down this effect, but this striking result is "consistent with the idea that infant attention to human stimuli may be biased toward the gender with which the infant has greater experience."

Is it a coincidence that one of the few attempts to document the effect of prior experience on infant behavior in a looking task reveals that experience matters? Of course not. Quinn's result is surprising to some only because it contradicts the guiding premise of nativism, namely, that even months of postnatal experience is unlikely to play a formative role in infant behavior.

Although experience may influence an infant's preference for male or female faces, the origin of the infant's preference for faces per se remains to be explained. As mentioned earlier, that a specific neurological insult can produce a selective deficit in face recognition cries out to many for a nativist interpretation. Indeed, the cries only get stronger when it is noted that newborns orient selectively toward geometric arrangements that have face-like properties.

Viewed abstractly, the human head and neck resembles a light bulb. Now, imagine a white light bulb against a black background, and imagine that we place three black dots inside the head of the bulb; one dot might represent the mouth and the other two dots might represent the eyes. Finally, imagine that we arrange the three dots to form an equilateral triangle with the two dots on top—so that it looks like an abstract face—or with the two dots on the bottom—so that it looks like nothing more than a triangle inside a light bulb. If you draw these arrangements for yourself, and if you are like most people, you will see a face in the former arrangement and a triangle in the latter; if you turn the page upside-down, your impressions will reverse. This sensitivity of face perception to the orientation of the page is called up-down asymmetry and indicates that we perceive faces as more than just a collection of features; critically, we are highly attuned to the configuration of the features, that is, the relationship among the parts. This is an important point to which we will return shortly.

As already mentioned, newborns look longer at three dots arranged to look like a face (∵) than at three dots in the upside-down arrangement (∴). But notice the bias in my presentation of this information. The three dots were simply arranged as triangles, one with the pointy end down (face) and one with the pointy end up (non-face). What if infants, instead of being sensitive to faces, actually prefer to look at objects that are top-heavy, regardless of whether the top-heavy arrangement looks like a face. For example, imagine a T-shaped object composed of five dots presented in the upright position (top-heavy) or the upside-down position (bottom-heavy). When newborns were presented with these kinds of stimuli, they preferred the top-heavy arrangement, just as they preferred the top-heavy face-like arrangements. These and other experiments are telling a new and unexpected story about face processing. One investigator, Chiara Turati, summarizes this story like this:

> [I]n order to explain the first steps in the development of specialization for face processing, it does not seem necessary to assume the existence of a specific innate cortical or subcortical bias toward faces. The earliest basis of face specialization appears to lie in the general functioning of the visual system, which constrains newborns to attend to certain broad classes of visual stimuli that include faces.

But how can we reconcile Turati's perspective with prosopagnosia, a selective deficit of face recognition that is typically caused by a small, discrete lesion in the right cerebral hemisphere?

In answering this question, we must first be clear on one critical point. Prosopagnosia is a deficit expressed by *adult* brains, not *infant* brains. In fact, the neural processing of faces by infants at six months of age is distributed across several regions on both sides of the brain and only begins to resemble the adult pattern of processing six months later. Annette Karmiloff-Smith acknowledges that damage to a normal adult brain can

> impair specialized areas of processing. But this does not necessarily mean that the brain started out with these specialized

circuits already in place. It could be that specialization builds up gradually and is actually the *product* of child development, not its starting point. So even if modules were identified in damaged adult brains, this in no way entails that they were prespecified in the newborn brain. . . . In other words, *domain specific outcomes do not necessarily entail domain-specific origins.*

Indeed, progressive sharpening of our perceptual abilities, reflected in increasing specialization of neural processing, may be a general feature of early development. For example, whereas human six-month-olds can distinguish between individual human *and* individual monkey faces, only human faces can be distinguished three months later; this interesting finding suggests that a relatively open face-recognition capability is narrowed during the first postnatal year as human infants gain experience with human faces. Analogously, human languages across the globe exhibit a diversity of sounds, only a subset of which are used in each individual language; between the ages of six and twelve months, infants become selectively tuned to the characteristic sounds produced within their native language and lose the ability to discriminate the characteristic sounds of non-native languages. Moreover, this same principle applies to language production, as some of us may know from trying to produce the trilled *r* of French or the characteristic sounds used in the clicking languages of East Africa.

We have now seen that damage to an adult brain, as in prosopagnosia, does not necessarily provide valuable information concerning the development of brain function. But are there developmental genetic disorders that disrupt brain function in such a way that we might learn something about the genetic basis of cognition?

Consider Williams Syndrome (WS), a developmental disorder with a known genetic basis in which numerical and problem-solving abilities are deficient but language processing and face recognition are not.

WS has been thought by some to target specific cognitive abilities via specific genetic pathways, sparing, for example, the innate "face recognition module." Rather than focus exclusively on the deficiencies of WS children and adults, however, Karmiloff-Smith and her colleagues were also intrigued by the possibility that seemingly normal behavior might be achieved in a novel way. For example, normal subjects and people with WS were asked to identify faces in the upright and inverted positions. Whereas normal subjects were better able to identify faces in the upright position, people with WS performed similarly when presented with either upright or inverted faces. How could this be? Apparently, people with WS explicitly memorize the individual features in a face, thereby making them insensitive to changes in face orientation. In contrast, normal subjects learn the configural relations among features, thereby making them sensitive to orientation; turn a face upside-down, and most of us can no longer identify it. This is but one example of how the same end-state can be achieved using different strategies that arise via different developmental pathways.

The nativist fixation on modules impels us to imagine that brain parts can be exchanged freely, like transistors in a radio. This is not, however, how brains function or develop. Yes, people with WS do indeed have abnormal brains. But as Karmiloff-Smith reminds us, "the abnormal brain is not a normal brain with parts intact and parts impaired. It is a brain that develops differently throughout embryogenesis and postnatal brain growth."

Nativism is alluring to scientists and the general public alike. It is a perspective that is represented prominently in a number of highly regarded psychology departments; many of the most recognizable developmental psychologists are nativists and their work is often published in highly visible journals and magazines. If developmental psychology were more akin to physics or chemistry, it would be difficult to argue that such acclaim and notoriety is undeserved. But developmental psychology, like all youthful sciences at one time or another, has been enveloped by a fad that gains its strength primarily from its easy distillation into morsels of information that are easily digestible by all who consume them. As we have seen throughout this book, however, development is a muddled, nonobvious, and complicated process. Attempts to explain this immense complexity through appeals to genes, instincts, and innateness simply do not do justice to how development actually happens.

Contrary to what nativists may think, the distinction between what is innate and what is learned is not particularly interesting; indeed, it has not been interesting for many years. Moreover, the more modern attempt to distinguish between knowledge that has a central origin versus knowledge that has a peripheral origin is yet another false dichotomy that does nothing to advance our understanding of knowledge or development. Removing the cachet of nativism requires first that we banish such dichotomies from our discourse.

Ethologists study complex behaviors that have transparent adaptive significance—for instance, filial and sexual imprinting—and in previous chapters I examined their developmental origins. In contrast, nativists focus on the simplest infant behaviors, not because they have adaptive significance—tongue protrusions and looking times do not ooze with adaptive significance—but for their use as winches to hoist complex competencies into the infant mind. It is in this sense that nativists ask leading questions and it is for this reason that the nativist inquiry is a dead-end.

For those on the lookout for human nature and who hoped that it would be found in the modest infant behaviors studied by developmental psychologists, the news is grim. This does not mean, however, that we have nothing to learn about human nature. We are, after all, a distinct species with many species-typical characteristics and behaviors.

So, if you are looking for human nature, consider that we crawl during infancy and walk upright thereafter, that we have opposable thumbs, that we lack fur, that we have a relatively large brain, that we typically give birth to one child at a time, that we are capable of producing speech in part because our larynx is positioned very low in our throat, that our bodies are uniquely shaped, that we have excellent vision and relatively poor olfaction, that we can only hold approximately seven items at once in working memory, and that we are adept at manipulating symbols and using language. Being human is the sum total of these and many other known characteristics as well as the unending cascades of interactions between them through developmental time. Attaining a satisfactory understanding of this complexity will take many years and a lot of hard work, but no one ever said it was going to be easy.

# STRAYING FROM THE HERD

"IT'S LIKE TRYING TO HOLD on to a greased pig!"

That is how one student of mine, in frustration, recently described his attempts to understand instinct. But how can such an ancient concept that describes behaviors of such obvious importance be the source of so much frustration? Unfortunately, numerous experts have made a bad situation worse, and there are many reasons for this: their attraction to grandiose theorizing, their distaste for mechanism, and more than a touch of intellectual laziness. This may sound harsh, but how else can one explain the incessant appeals, by scientists who should know better, to genetic programs, blueprints, and recipes; to innate neural modules that govern the expression of everything from jealousy to communication-pragmatics; to human newborns that emerge from the womb with a knowledge of physics and arithmetic, as well as the ability to imitate and even read the minds of others. Of course, every field is populated with individuals who have committed blunders. But these are doozies, repeated over and over again, monotonously, like a child fixated on a new favorite phrase.

Ironically, nativists have expelled development from developmental psychology at a time when evolutionary biology is inviting development back in. For example, with *Developmental Plasticity and Evolution,* Mary Jane West-Eberhard has written an imposing and impressive book that should help to establish the notion that development is the leading edge of evolution, not the peripheral diversion that so many have thought. Among the central themes of her book is that the true promise of evolutionary theory will only be attained when we fully appreciate how the environment effects evolutionary change by modifying developmental processes. Similar to Gottlieb and his notion of the developmental manifold, West-Eberhard envisions a "responsive phenotype at the center of development" for which it "is of little developmental consequence" whether change occurs via genetic or non-genetic causes. This simple observation, she notes, sounds the "death knell of the nature-nurture controversy, for it puts genes in perspective without detracting from their importance."

Central to West-Eberhard's approach to understanding the links between development and evolution is the need to focus on the actual mechanisms that, moment to moment, underlie these processes. All too often in the recent history of evolutionary biology, the favoring of function over mechanism, or, of ultimate over proximate causation, has turned evolutionary thinking into an idealized parlor game where cards and chips don't matter. But this is a perspective that seems to be changing. For her part, West-Eberhard sums up the concerns of many of us with a simple warning that, if heeded, would prevent many needless arguments about instincts, genes, and other troublesome concepts: "Confused concepts are virtually inevitable if data on real mechanisms are ignored. . . ."

Real mechanisms do not include imaginary genes that tell us what to name our dog or what color car we should purchase. And real mechanisms do not include imaginary neural modules that innately determine our ability to express grammar or understand physics. Unfortunately, many developmental and evolutionary psychologists have embraced these illusions and we have seen the results. Every time that we have closely examined their claims, we have found faulty experiments, far-fetched interpretations, or both. In short, nativists and evolutionary psychologists have draped themselves in the blanket of science, but, when all is said and done, they are merely telling bedtime stories for adults.

The imaginary genes and modules of nativists and evolutionary psychologists do not work as explanatory devices in part because there exists a sizable gap between genes and modules and the phenomena that they are invoked to explain. In Gunther Stent's words, there are *too many removes* between a gene and the act of naming a dog, or between an innate neural module and the expression of grammar. The goal of our science should be to bridge that gap, not blindly disregard its existence. Action at a distance is not a property of gravity (Albert Einstein considered any such action "spooky" and took care of the problem with his General Theory of Relativity), and it certainly isn't a a property of genes.

The fixation on genes and modules with extraordinary powers reflects still another lingering, pernicious problem: the belief that every complex system must have a single, essential locus of control. Thus, many view development as controlled by genes, and behavior as controlled by neural modules. To appreciate the pitfalls of this fixation on control, you may remember a scene from the beginning of the movie *Superman* in which Lois Lane is a passenger in a helicopter when it crashes on the roof of a skyscraper. As she falls to a seemingly certain death, Superman arrives in time to deftly catch her and carry her to safety. "Easy, Miss, I've got you," he says nonchalantly, to which Lois replies frantically, "You've got *me*? Who's got *you*?" And so it is with high-impact, ungrounded words like *gene, module,* and especially *instinct.* They appear to resolve questions of control until we think to ask what is controlling them.

The search for the ultimate locus of control, then, is often futile, especially when we are considering complex systems like organisms. We saw in Chapter 2 how the argument from design misleads us into thinking that the very existence of complexity demands the existence of a designer. A similar trap concerns what computer scientist Michael Resnick calls the *centralized mindset,* which he illustrates using the following example:

> A flock of birds sweeps across the sky. Like a well-choreographed dance troupe, the birds veer to the left in unison. Then, suddenly, they all dart to the right and swoop down toward the ground. Each movement seems perfectly coordinated. The flock as a whole is as graceful—maybe more graceful—than any of the birds within it.

The centralized mindset is revealed when we are asked to identify the source of the flock's synchronous behavior and, in response, we reflexively imagine that there must exist one bird that, like an army general, leads all the others. But this is not how a flock works. As Resnick points out, each bird "follows a set of simple rules, reacting to the movements of the birds nearby. . . . Orderly flock patterns arise from these simple, local interactions." Thus, in flocks of birds and other self-organizing systems—such as bee swarms, the weather, gene networks, and populations of neurons—control is nowhere and everywhere depending on how you look at it. Moreover, as if written by an epigeneticist rather than a computer scientist, Resnick notes that a "richer view of the environment is particularly important in thinking about decentralized and self-organizing systems."

Thus, any comprehensive account of behavioral development must accommodate all of the cascading and intersecting influences that intervene between genes and behavior. In one such account, Tim Johnston and Laura Edwards break down the broad classes of inputs to behavior as follows:

*Genetic activity:* refers to the "transcription of DNA that underlies the involvement of the genes both in protein synthesis and in the regulation of their own activity";

*Sensory stimulation:* refers to "any influence that acts through the developing animal's sense organs and is processed by its nervous system, including all the effects of learning and of experience more generally construed"; and

*Physical influences:* refers to "all other environmental effects (both physical and chemical), including diet, temperature, pH, salinity, gravity, and the mechanical stresses exerted during movement."

Starting with these three broad classes of inputs, and factoring in the intricacies of developmental timing, the complexities increase rapidly. For example, nearly all influences within the system are bidirectional, including those on behavior; thus, sensory stimulation and physical influences affect behavior, and behavior in turn affects sensory stimulation and physical influences. Behavior even has an indirect effect on gene activity (the effect must be indirect because DNA is securely ensconced within the cell membrane, far beyond the immediate reach of behavior). For example, everyday experiences alter sensory stimulation and subsequently induce changes in neural activity; this neural activity initiates a cascade of events that, within minutes, results in altered gene activity; moreover, the proteins that result from this altered gene activity can further influence gene expression in the brain. Of course, all of these influences conspire to shape the developing animal, their relative influence waxing and waning on a moment-to-moment basis. This account of development may seem monstrously complex, but what is truly monstrous is the notion that animals—whether a worm, dog, or human—could ever be adequately explained using the simple-minded models and metaphors that are still invoked to explain behavioral development.

We have come a long way since Chapter 1 and the discussion of herding behavior and other complex behaviors by dogs. It is now time to return to this topic with a greater appreciation for the notion that complex behaviors do not simply happen. They develop.

## *Domestic Help*

Charles Darwin recognized the value of domesticated species for furthering his understanding of evolution and behavior, and he spent many hours in the company of "pigeon fanciers" to learn more about the strange behaviors of this once popular pet. As he wrote in *The Origin of Species*:

> Domestic instincts are sometimes spoken of as actions which have become inherited solely from long-continued and compulsory habit; but this is not true. No one would ever have thought of teaching, or probably could have taught, the tumbler-pigeon to tumble—an action which, as I have witnessed, is performed by young birds, that have never seen a pigeon tumble. We may believe that some one pigeon showed a slight tendency to this strange habit, and that the long-continued selection of the best individuals in successive generations made tumblers what they now are; and near Glasgow there are house-tumblers, as I hear from Mr. Brent, which cannot fly eighteen inches high without going head over heels.

Darwin then applies this same argument to the ultimate of all domesticated species, the dog:

> It may be doubted whether any one would have thought of training a dog to point, had not some one dog naturally shown a tendency in this line; and this is known occasionally to happen, as I once saw, in a pure terrier. . . .

I made a similar observation several months ago as I watched Katy, our Border collie mix, bounding across the yard; suddenly, for no reason that I could discern, she stopped dead in her tracks, tail high, and lifted one of her paws off of the ground to produce the three-point stance often associated with pointing behavior. I have now seen her do this many times, so it has apparently become part of her behavioral repertoire. I was, naturally, intrigued.

Pointing is typical of many dog breeds (although perhaps not universally expressed by all modern breeds), so it is not surprising that Katy would occasionally exhibit that behavior. But pointing is only one among many behaviors that comprise the attributes of a good pointing dog. In fact, first-class pointing dogs must be carefully trained to become useful companions during a hunt; for example, they must learn to remain steady in the face of a deafening shotgun blast and must resist the urge to chomp down on a succulent bird during retrieval.

But it all starts with one basic behavior—the point—and it is on this behavior that breeders focus when analyzing the potential of a young puppy. In Harold Adams's and Dave Meisner's video, *Point! Refining the Pointing Instinct,* we are shown what breeders look for in a puppy. The assessment technique is simple: a bird feather is attached by a string to the end of a stick, and the feather is dangled on the ground. Even at only four months of age, a dog with the potential to become a good pointer will orient rapidly toward the feather, tense its body, and curl its tail upward. The narrator of the video speaks of a hard, stylish, intense point, and comments that a puppy that possesses these qualities has "the right stuff to start with," that is, the pointing instinct. And this start is essential. The narrator explains, "This tape will not teach you how to train your dog to point because that's not possible. Pointing is genetic; it's an inherited instinct, the result of generations of selective breeding."

Pointing is as good a candidate for an instinct as any behavior. It is obviously inherited in that it is a distinct behavior of dogs—and not, for example, cows—and can be enhanced through selective breeding. Also, it is a species-typical behavior that is likely part of the dog's predatory behavioral sequence; indeed, Darwin believed that pointing is probably "only the exaggerated pause of an animal preparing to spring on its prey." But as for the genetic and experiential mechanisms that lead to the expression of pointing behavior, we know virtually nothing. Clearly, the narrator's claim that pointing is "genetic" is based solely on the belief that any "inherited instinct" must also be genetically determined. But we have seen that this belief is mistaken: inheritance does not necessarily implicate genes, and it certainly does not imply genetic determination (unless all one means by this is that genes are *somehow* involved, which is trivially true of every behavioral trait).

Of course, genetic factors may very well play an important role in the propensity of some dogs to point more readily or intensely than others. And with recent investments in the so-called Dog Genome Project, the search for such genes may not be too far away. But we need to remember that the identification of genes that influence a behavior is only one small step toward understanding the complex process of behavioral development.

The Dog Genome Project is about far more than just behavior; in fact, dogs offer enormous opportunities to geneticists interested in inherited human diseases. One reason for this is that it is relatively easy to identify genetic markers of disease in inbred populations; and dogs, with over three hundred contemporary breeds, are probably the most intensively inbred species on the planet. In addition, dog breeders and the organizations with which they affiliate—such as the American Kennel Club—work vigorously to preserve the "purity" of each breed by controlling mating opportunities and maintaining meticulous genealogical records; indeed, the breed barrier rule specifies that registration of a dog as a member of a particular breed requires that both its parents are also registered members. With all of this inbreeding and the consequent increased chances for the expression of unwanted recessive traits, numerous "genetic" disorders have been identified—360 and counting—that often resemble disorders found in humans.

One of the most stunning accomplishments of canine genetic research is related to the sleep disorder narcolepsy, which is characterized most famously by the sudden onset of sleep during the day. Victims of this disorder can be normally conversant one moment and laid out on the floor the next, completely paralyzed and yet aware of what is going on around them. The discovery over twenty-five years ago that some Doberman Pinschers exhibit the classic signs of narcolepsy was remarkable; it was also a little ironic that a dog, bred in the late nineteenth century as personal guardians and renowned for their capacity for ferocity, would contribute to our understanding of a disorder that renders its victim harmless when it gets a little excited. Nonetheless, in a flurry of scientific advances over the last seven years, researchers identified and cloned the mutant gene that codes for the receptor for a newly discovered neurotransmitter, called orexin, that is now known to play a role in the neural circuit that underlies the sleep-wake cycle. Although human narcolepsy appears to involve this same neural circuit, it is not yet clear that human and canine narcolepsy have the same genetic basis. Nor should we expect them to: there are many possible genetic and extra-genetic pathways through which narcolepsy, and many other disorders, can be produced.

The blazing pace of narcolepsy research reflects a field that was ripe for rapid progress. Caution is needed, however, when dealing with lesser-known disorders, and dog geneticists are aware that breed-specific disorders do not necessarily imply specific genetic contributions. For example, Elaine Ostrander and Leonid Kruglyak raise the cautionary tale of the Shar-Pei. Bred in China for two millennia as herders and easily recognized by their loose, wrinkly skin, these dogs suffer more than do dogs of other breeds from infectious diseases of the eyes and skin. Were geneticists to search for a breed-specific genetic basis for the Shar-Pei's increased susceptibility to infection, they would probably find one. But such a discovery would not indicate the presence of a "disease gene" or a mutated immune system. Why? Because the higher rates of infection in this breed are "the result of physical characteristics associated with the breed standard," that is, the difficulties associated with keeping the wrinkly folds of the Shar-Pei's skin clean. Indeed, many canine afflictions may be due to the imposition of arbitrary breed standards that place unreasonable physical and physiological stresses on these animals.

As difficult as diseases are to understand, behaviors pose even more difficulties because so many more systems are involved, including, of course, the brain. The following passage, by science writer Stephen Budiansky, gives a sense of the opportunities afforded by dogs for the study of behavior while simultaneously illustrating some potential pitfalls:

> Other studies have turned up some remarkably narrow and distinctive behavioral lineages that further demonstrate the extent to which canine behavior is genetically determined. Certain strains of Siberian huskies and pointers have developed a strongly inherited shyness or aversion to human beings; when kept under identical conditions in identical kennels, the shy dogs will stay back (or, in the case of the pointers, actually freeze and quiver when people approach), while the normal dogs come up to be petted. Breeders have succeeded in producing lines of bloodhounds that bark or do not bark while trailing a scent; of Dalmatians that do or do not take up the proper "coaching" position, trotting under the front axle of a carriage, very close to the heels of the horses; and even of miniature poodles that do or do not "shake hands."

These canine examples cover a wide range of behavioral categories, yet they are all lumped together as illustrative of genetic determination on the sole basis that they occur in "distinctive behavioral lineages." The earlier example of the Shar-Pei should provide an antidote to such thinking, as should the many other examples that we have reviewed in this book. For example, if we observed a duckling approaching its mother while she emits her assembly call, we might be tempted to assume that the duckling's attraction to its mother's call is genetically determined; but, as we now know from the work of Gilbert Gottlieb, we would be mistaken. *The fact is that no one has yet to examine behavioral development in dogs with the same methodological rigor as Gottlieb has in ducklings.*

Consider the example of miniature poodles that do not shake hands. What might be the cause of such a specific behavioral deficit? If, as Budiansky suggests, this trait is genetically determined, we might imagine that there is a "hand-shake neural module" that is present in some miniature poodles and absent in others. But before we go down this road, should we not consider the possibility that there is a more simple and direct explanation that connects this odd observation in miniature poodles with what is known about similar behaviors in other animals?

To begin, recall Konrad Lorenz's confident assertion that overwing head-scratching in birds is part of their genetic heritage. As it turned out, subsequent research demonstrated that head-scratching is much more responsive to context than Lorenz's conception would allow; for example, some birds scratch differently while perching than while flying. Based on all of the available evidence, Burtt and Hailman concluded that posture, balance, and center of gravity play a pivotal role in the expression of this behavior. Next, consider the fact that mice are unable to groom themselves with their forepaws when newly born and only develop this ability over the next ten days. But before we conclude that a neural "grooming module" is maturing over this period, we should realize that a newborn mouse can exhibit grooming-like movements with its forepaws if we prop it up in an upright position. In other words, providing postural support to a newborn mouse frees the forepaws to engage in other behaviors, such as grooming.

Head-scratching in birds and grooming in mice suggest an answer to the miniature poodle problem: Specifically, some miniature poodles may refrain from shaking hands for biomechanical, not neural, reasons. Thus, it could be that selective breeding for body size, body shape, weight distribution between the front and back halves of the body, or relative limb strength might alter the ability of some miniature poodles to support themselves on three limbs and shake with the fourth. Moreover, this problem may be typical of other miniature breeds and might be overcome by providing postural support to small dogs while training them to shake. This example may seem trivial but the lesson is important: we gain nothing by attributing a behavioral trait to genes when we know little about the processes that influence the expression of that behavior.

If we keep in mind the model of Johnston and Edwards described above, we will learn to expect gene-centered approaches to be enlightening at times—as in the case of narcoleptic Doberman Pinschers—and distracting at others—as in the case of wrinkly Shar-Peis and non-shaking miniature poodles. Again, gene activity, sensory stimulation, and physical influences are all important factors in behavioral development and our goal should be to unravel the contributions of each. When we lose sight of one, we blind ourselves to the true complexity of development.

## *Back to Basics*

Watching a well-trained pointing dog during a hunt can easily lead an observer to conclude that pointing is a uniquely canine behavior. On the contrary, the core of the pointing dog's usefulness to hunters rests on a basic orienting response that is unique neither to pointers nor to other dogs. In fact, every mammal and bird orients to novel or relevant stimuli in a way that betrays the sensory system that is being engaged: when owls hear a mouse scurrying across the ground, they turn their entire head so that their eyes, which have limited mobility within their sockets, can focus upon their prey; rabbits rotate their ear pinna, often one at a time, to locate the source of a sound; and dogs stiffen their bodies when they sniff vigorously, a behavior that looks to us like a point. Of course, dogs are not pointing in the same way that we do—they are simply orienting toward the odor. And we have learned to take advantage of the bodily information that the dogs provide.

Orienting responses have the characteristics that we commonly associate with reflexes: clap your hands loudly to the side of a friend's head and watch how his head whips around. The behavior of pointing dogs, then, is relatively easy to comprehend: it comprises a basic orienting response to odors with a lot of training piled on top. What, then, is the difference between an instinct and a reflex—between a point, a peck, and falling asleep in the middle of the day? As discussed in Chapter 5, this was exactly the argument that Pavlov was making when he searched in vain for "any line of demarcation" between the two. As we peel away the complexity of a behavior to reveal its components and developmental origins, labels that were once seemingly infused with so much meaning fade like the campaign slogans of a bygone era.

Whereas some might think that pointing and the other behaviors of hunting dogs—sculpted by breeders for centuries and expressed by puppies at a very young age—would be the product of simple genetic rules, this is not the case. In fact, breeders have found that superior pointing dogs do not give rise to superior pointing progeny. But even more surprising is that expression by offspring of those traits considered most important in a useful hunting dog are predicted best by the identity of the mother, not the father. In other words, the mother appears to contribute predominantly to the very differences in dog behavior that lead so many to focus on genes.

We have seen in this book numerous examples of how maternal contributions to development can be provided through a variety of nongenetic sources: cytoplasm in the fertilized egg, diet-induced methylation in utero, and maternal behavior. Consider the role of maternal behavior: like rat pups, dog puppies must be licked in the anogenital region by the mother in order to urinate and defecate. So, might behavioral differences among canine mothers influence the expression of breed-specific behaviors, just as was seen in Chapter 3 that they influence fearfulness and spatial learning in rats? Isn't it possible, perhaps even likely, that breed-specific behaviors are inherited in part due to breed-specific aspects of the early environment, including specific attributes of the mother and her behavior? As far as I know, such questions have never been addressed in dogs.

Much of what we do know about early development in dogs we owe to the classic work of John Paul Scott and his collaborators, beginning over fifty years ago. Scott provided an important framework for conceptualizing early development that complemented similar work being done at the time by developmental psychobiologists on cats and rats. This framework emphasized particular stages of development during which important milestones are reached. (Interestingly, Scott did not recognize the importance of the prenatal period for behavioral development in dogs.) This work, especially concerning the socialization process, has shaped the practices of breeders concerning when to wean a puppy from its mother and when to begin to socialize dogs with humans so as to ensure the development of a healthy pet. For example, overcoming a puppy's initial fear of humans was thought to occur most rapidly between the ages of three and five weeks and the development of a close social relationship was thought to occur most readily between six and eight weeks. Although recent evidence suggests that the now-standard practice of removing puppies from their mother's care at six weeks is detrimental to their subsequent health, and that prolonged maternal care through twelve weeks is preferable, the basic point is indisputable: Early experiences matter for the development of a well-adjusted, healthy dog.

But does early experience matter for the expression of breed-specific instinctive behaviors? Consider livestock guarding dogs, like the Great Pyrenees, that accompany sheep and other livestock during the day, sit beside them, and protect them from predators without themselves showing any inclination to stalk, chase, bite, and eat them. Although some might jump to the conclusion that this failure to display the predatory behaviors of the wolf ancestor means that guarding dogs have had their natural predatory instincts bred out of them, such a conclusion would seem to overlook a simple fact: the most effective guarding dogs are those that have, ideally before the age of eight weeks, been reared with livestock! In other words, the expression of the guarding dog "instinct" is not inborn or genetically determined; rather, this behavior depends on the early socialization of the dogs with livestock. In a similar vein, the developmental psychobiologist Zing-Yang Kuo reported in 1930 that kittens will form social attachments with rats if they are reared with them and, as mature cats, will refrain from exhibiting their cat-typical rat-killing tendencies. All of us really can get along.

Thus far, we have made some headway toward understanding the basics of pointing and guarding. But what about herding? When we left this issue in Chapter 1, there were numerous herding-related behaviors that begged for an explanation. You may recall the herding certification test, with independent measures for style, approach, eye, wearing, bark, grouping, power, temperament, and interest. But what do we actually know about these measures? Are these truly independent measures of herding behavior, or do dogs that score high, for example, on style also tend to score high on one or more of the other measures? Do any of these measures depend on early experience or explicit training?

As with pointing dogs, herding dogs receive extensive training. In fact, Border collies that compete in herding trials are typically taught dozens of whistle and voice commands. So, as with pointers and their powerful orienting response, we should expect Border collies to readily exhibit certain behaviors that increase their value as herders relative to other breeds. Can we identify a behavior that might fit the bill?

Having lived with Katy for over two years, there is a characteristic that sets her apart from any other dog that I have known. Once Katy fixes her eyes on a moving object, she doesn't let go. This unwavering stare, called *showing eye,* is considered one of the important building blocks of herding behavior. Melissa Fleming, working with Elaine Ostrander, targeted this behavior as a potential candidate to study the genetic contributions to dog behavior. Fleming chose to compare Border collies with Newfoundlands, large dogs that do not readily herd but are known instead for their attraction to water (and thus their usefulness as water rescue dogs). In one test, the staring responses of these dogs to a remote-controlled car was assessed for two minutes. Whereas the Border collies fixated on the car for the entire two-minute test, the Newfoundlands stared on average for only ten seconds. Here, then, is a quantifiable difference between two breeds that seems to be related to clear breed-specific differences in their herding tendencies.

Of course, there is much more to herding than just staring, but the fact is that we simply do not know very much about the development of showing eye in Border collies (or other breeds) or how the development of this behavior influences (or is influenced by) the development of other herding-related behaviors, such as stalking and chasing. Thus, if we truly want to move beyond clichés toward an understanding of these complex behaviors, we need a comprehensive approach that considers the full range of epigenetic factors that produce a herding dog.

Over and over again, we have encountered the idea that early sensory and perceptual biases or predispositions can play formative roles in shaping the development of complex behavior, from pecking in laughing gulls to face recognition in human infants. Here again, now with dogs, we see how behaviors that seem incomprehensibly complex can be broken down into component parts to reveal relatively simple foundations. But now we confront a new dilemma: How could breeders, absent the knowledge of the building blocks of instinctive behavior, ever manage to produce useful working dogs? Presumably we do not believe that breeders were able to identify each of the critical components of an instinctive behavior, including perceptual biases and reflexes, and breed selectively for each of them. But if that's not how they did it, then what explanation are we left with?

### *Forests and Trees*

As a first step toward answering this last question, imagine Gulliver being escorted by the Lilliputians to a forest filled with trees of varying heights. For their own silly little reasons, the Lilliputians feel that the forest has grown too tall and so they ask Gulliver to reduce the height of the forest by half. Obliging his hosts, Gulliver measures the height of the tallest tree (which does not exceed his own height), divides by two and, using his ax, chops down all of the trees that exceed the limit. After a hard day's work, the Lilliputians inspect the result in horror. Because of their diminutive size, the Lilliputians never realized that the forest was composed of two kinds of trees: A highly prized tree that grows to great heights, and a tree of little value and even smaller stature. By asking Gulliver to reduce the height of the forest, they had unwittingly asked him to decimate the only trees of value within it.

Indeed, we often choose products and lifestyles based predominantly on one attractive feature without seeing in advance what else might come along for the ride. Thus, we might adopt a high-protein diet but find that our cholesterol level shoots up as we consume more meat; we might move out to the country for the fresh air only to be undone by the allergens; or we might opt for the apparent safety of an SUV but be overwhelmed by the exorbitant costs associated with keeping these behemoths filled with gasoline. For these decisions, however, the links between cause and effect are transparent. When we make similar decisions in the breeding of animals, the results are anything but transparent.

In the late 1950s, working in the isolated Siberian city of Novosibirsk, Dmitry Belyaev began an experiment that was to last for over forty years. A geneticist by training, Belyaev was interested in recreating the process of domestication using a species that was closely related to the dog but that had never been domesticated. He chose the silver fox. By 1999, fifteen years after Belyaev's death, he and his colleagues had bred forty-five thousand foxes over more than thirty generations to produce a tiny group of one hundred friendly foxes. As Lyudmila Trut describes the members of this "domesticated elite," they wag their tails, whine, and "are eager to establish human contact, whimpering to attract attention and sniffing and licking experimenters like dogs."

The production of docile animals from often vicious forebears was not the only result of this unprecedented breeding program. The process of domestication brought with it a variety of additional features, many of which we associate with domestication in other species, including dogs, horses, and cattle: erect ears became floppy; tails curled upward; and coat color changed from uniformly pigmented to piebald, that is, a combination of pigmented and unpigmented areas (the black and white coloration of some Border collies is one example of this). There were many other changes as well: the domesticated elites have smaller heads and shortened snouts, they reach sexual maturity about one month earlier, they give birth to larger litters, and they exhibit differences in hormone production and brain neurochemistry.

How did Belyaev and his colleagues manage to induce all of these changes—in anatomy, physiology, and behavior—in just forty years? Did they only breed those foxes with smaller heads, floppy ears, curled tails, or unpigmented skin? *No.* Did they only breed those foxes that reached sexual maturity quickly? *No.* Did they train their foxes in some unique way or raise them in the presence of humans or dogs? *No.* All they did, in fact, was give young foxes a monthly test: "When a pup is one month old, an experimenter offers it food from his hand while trying to stroke and handle the pup. . . . The test is repeated monthly until the pups are six or seven months old." Then, the pups are given a "tameness" score and only those foxes that score high on this score are allowed to breed. Thus, selection on the basis of this single behavioral dimension is sufficient to produce the amalgam of changes already described, resulting in the creation of a domesticated fox.

Now here is the big question: How can selective breeding of only the tamest animals lead to such a diversity of anatomical, physiological, and behavioral changes? After all, there is no rational connection between tameness and a small head, or a curled tail, or unpigmented skin. Nonetheless, all of these features are clearly connected. But what is it that connects them?

It turns out that by selecting for "tamability" Belyaev was selecting for foxes that developed at different rates than those in the original founder population from the 1950s. Belyaev was, in Gilbert Gottlieb's words, selecting for the *entire developmental manifold.* And in this case, the domesticated elite seemed to develop more slowly. Delayed sexual maturity is one aspect of this retarded development, but there were other signs as well; for example, the domesticated elite showed delayed fear responses that emerged at nine weeks of age, compared to six weeks in their nondomesticated cousins.

But the list of developmental changes does not end there. In fact, nearly all of the identifiable changes in the domesticated elites can be attributed to changes in the timing of development, or *heterochrony.* Developmental timing is a fundamental mechanism of evolutionary change; if one compares the early embryonic stages of a human and a chicken, it becomes readily apparent that the dramatic differences between adult humans and chickens arise in the course of development and are not established at the outset. For a more transparent example, consider an adult wolf—with its sharp, long snout and erect ears—and compare it to a wolf puppy—with its rounded, short snout and droopy ears—and you immediately get the picture: adult domesticated dogs bear a striking resemblance to young wolves. It is as if the process of domestication is one of developmental retardation, of producing juvenile versions of ancestral adults.

Belyaev's selection technique likely produced changes in the genetic make-up of the domesticated elites. But just as important is the fact that Belyaev, merely by focusing on tamability, unwittingly restructured every fox feature that was sensitive to changes in the *when* of development. As already mentioned, shortened snouts and floppy ears are puppy-like features; and delayed sexual maturity fits in nicely with the general concept of developmental delay. But changing the timing of development also changes the complex interplay between developmental events, thereby producing unexpected results. For example, the piebald coat color of domesticated species likely arises from alterations in the complex interplay between the embryonic cells responsible for the pigmentation of fur. In addition, a delay in the developmental onset of fear responses leaves open the window for socialization of puppies with other animals, including humans.

The delayed fear responding of the domesticated elite is particularly interesting in light of the research on rats by Michael Meaney and his colleagues, discussed in Chapter 4, on the modulating role played by maternal care, including anogenital licking, on the fear responses of offspring. You may recall, for example, that female rats raised by mothers that engage in a lot of anogenital licking exhibit reduced fearfulness as adults and also display the same mothering style that they experienced as infants. In light of these findings, is it possible that the selective breeding of tamable silver foxes resulted in the selection of different styles of maternal care, and that maternal style might account in part for the behavioral changes witnessed over the course of Belyaev's experiment? We don't yet know because no one has looked.

Whether Belyaev's model of domestication in the fox provides the full explanation for domestication in other species, including dogs, is not known at this time. What we do know, however, is that selective breeding is a complicated affair that entails complex, largely unexplored alterations in gene expression, developmental timing, and early experience. Thus, although we do not yet understand the origins and mechanisms of the remarkable herding, guarding, and hunting behaviors of dogs, we now have sufficient information to know where and how to look.

## *Homeward Bound*

The many meanings of instinct provide a roadmap to the wrong turns and dead ends that have hindered progress in our understanding of behavior and behavioral development. Recall from Chapter 1 that Patrick Bateson lists several such meanings, including present at birth, not learned, developed prior to use, fixed, species-typical, served by a specific neural module, produced through the evolutionary process, and genetically determined. These meanings, many of which bear no necessary relation to the others, are like conceptual parasites that have overwhelmed the host. Overburdened by unnecessary obsessions with innateness, fixity, modularity, and determination, the instinct concept—like the gene concept—may have become an impediment to further understanding.

Thus, from its beginnings as an explanation for the seemingly rational behaviors of nonhuman animals, the instinct concept has become a universal implement for scratching every nativist itch. Resisting the urge to scratch requires something more than willpower: it requires a fundamental reworking of our notion of inheritance. As the instinct concept has been expanded to encompass diverse aspects of our nature, the meaning of biological inheritance has been constricted to mean little more than what is passed down through our genes. But once we loosen the restraints on what is meant by inheritance—restraints that were promoted by mid-twentieth century biologists determined to forge an exclusive link between evolutionary change and genetic change—we see that a portal is opened that permits reentry to development as a prominent player in the central questions of biology and behavior. In other words, when we truly appreciate the multiple dimensions of inheritance, we can no longer ignore how development actually happens.

Reworking our notion of inheritance begins with the simple, irrefutable fact that the fertilized egg, in its entirety, is inherited from our parents: this includes the DNA-containing nucleus, yes, but also the full complement of nongenetic, cytoplasmic factors that are passed down from the mother and that critically modulate the expression of the genes through local, mechanistic interactions. And contained within this expanded notion of inheritance lies a perspective that compels a more realistic, dynamic, and contextualized view of development. This perspective applies at every stage of development and is captured by such phrases, encountered earlier, as *species-typical environment, normally occurring experience,* and the *normal arrangement of the animal's world.* Therefore, cytoplasmic factors are only the first in a long and continuous series of inherited environmental and experiential factors that reliably shape development from generation to generation. It is always out of this process that complex behaviors—instincts—emerge.

Armed with a recalibrated notion of inheritance and an expanded appreciation of development, the true nature of instinctive behavior reveals itself. The nascent instinct is little more than a simple reaction—a peck, a stare, an orientation response—that is biased toward particular features in the environment. This reaction is elaborated during development through species-typical and individual experiences, which may include the influence of the physical environment and sensory stimulation. Experience can also include explicit training by a member of the same species (as in song learning in cowbirds) or by a different species (as in the training of dogs by humans). Regardless, over time, the origins of behavior—the reflexes, predispositions, and species-typical and individual experiences—fade into the past, leaving what *appears* to be a finished product that is predetermined, rational, and designed. It is as if time throws a blanket over complexity, smoothing out the uneven terrain and feeding the illusion that we have found a comfortable place to lie down and nap.

There have been many changes in our household since I began this book. Sadly, at fifteen years of age, our two Bichons died, leaving Katy alone without her herd. To be sure, Katy is a year older and calmer, but the striking change in her demeanor with the passing of the second Bichon goes far beyond age. Now she has no one to chase in the yard and no one to prevent from ascending or descending the stairs. And because the targets of Katy's mischief have disappeared, we no longer find ourselves yelling at her day and night. Like an alcoholic in a world without alcohol, Katy simply no longer has the opportunity to exhibit her destructive behaviors. She has become, literally overnight, a perfect dog.

Same DNA. Different dog.

# NOTES

These notes provide source information for all quoted passages as well as for those facts deemed important but whose sources may not be apparent in the text. Complete references can be found in the bibliography.

EPIGRAPHS

p. ix   "What then, once again, is instinct?": S. Diamond, 1974, p. 243.

p. ix   "Nothing is easier": R. J. Richards, 1987, p. 24.

INTRODUCTION

p. xi   "Many instincts are so wonderful": Darwin, 1859/1983, p. 317.

p. xii  "An action which we ourselves": Ibid., pp. 317-318.

CHAPTER 1: A HERD MENTALITY

p. 5    Gray wolves began living in close proximity with our human ancestors as many as 400,000 years ago: Clutton-Brock, 1995.

p. 5    Dogs diverged from wolves as many as 135,000 years ago: see Wayne & Ostrander, 1999.

p. 6    Herding dogs are in turn subdivided: Holland, 1994.

p. 6    These behaviors include: Coppinger & Schneider, 1995, p. 30.

p. 7    Virgil Holland asserts: Holland, 1994, pp. 3-4.

p. 7    John Holmes defends the belief: Holmes, 1998, pp. 49-50.

p. 8    To become certified: e.g., see http://www.glassportal.com/herding/instincts.htm

p. 8     "suddenly has the light go on": Ibid.

p. 9     "It appears that dogs are genetically programmed": Coppinger & Schneider, 1995, pp. 31-32.

p. 10    "Apart from its colloquial uses": Bateson, 2002, p. 2212.

p. 11    Morgan appreciated how the same word: Morgan, 1896/1973, pp. 2-3.

p. 11    "rediscovered the malodorous aspects of instinct": Bateson, 2000, p. 190.

p. 12    "behavior based on inborn neural circuits": Gould & Gould, 1999, p. 24.

p. 12    Elsewhere, they state that: Gould & Gould, 1999, p. 9.

p. 14    the ability of a human newborn to recognize her mother's voice develops within the womb: DeCasper & Fifer, 1980.

p. 16    maternal instinct of female rats: see Lehrman, 1953, p. 342.

p. 16    profound fear of snakes exhibited by Japanese macaques: Masataka, 1993.

p. 17    newly hatched chicks that are prevented from seeing see their toes: Wallman, 1979.

## CHAPTER 2: DESIGNER THINKING

p. 20    Long before Darwin: For a review of early attempts to understand instinct, see Diamond, 1973.

p. 21    Knowing the etymology of a word: Ibid., p. 151.

p. 23    "Consider, anatomize the eye": Hume, 1776/1985, p. 25.

p. 23    "To suppose that the eye": Darwin, 1859/1983, p. 227.

p. 24    "numerous gradations from a simple and imperfect eye": Ibid., p. 228.

p. 24    "inaccurate workmanship . . . of the great machine of nature": Hume, 1776/1985, p. 73.

p. 24    "was only the first rude essay": Ibid., p. 37.

p. 24    "Odd arrangements and funny solutions are the proof of evolution": Gould, 1982, p. 21.

p. 25    "And if we are not contented": Hume, 1776/1985, p. 80.

p. 26    one anecdote recounted by Romanes: Romanes, 1883/1977, pp. 364-365.

p. 27    "If a hand can do it, why not a paw?": Ibid., p. 422.

p. 29    Consider a dining room table: Petroski, 1992a, pp. 3-21.

p. 32    artifacts "do not spring fully formed": Ibid., pp. 19-20.

## CHAPTER 3: SPOOKY

p. 35    With great fanfare in mid-February 2001: Lander et al., 2001; Venter et al., 2001.

p. 36    "Unfortunately," according to Collins: Collins et al., 2001, p. 27.

p. 37    "more than nature": Ibid., p. 29.

p. 37    "a kind of mentality": Keller, 2000, p. 47.

p. 37    "the cell's brain": Baltimore, 1984.

p. 37    "in helping to reveal the naëveté of those hopes": Keller, 2000, p. 70.

p. 38    "a past that has been documented by many others": for example, see Lewontin et al., 1984.

p. 38    story of twin girls: Gootman, 2003.

p. 39    "despite their well-established shortcomings": for example, see Wahlsten, 2000.

p. 40    "described by others": for example, see Joseph, 2001.

p. 40    "Many people are skeptical of such anecdotes": Pinker, 1994, pp. 327-328.

p. 40    "We are awed by these similarities": Hauser, 2000, p. 111.

p. 41    Consider this college-age pair and the striking similarities between them: see Joseph, 2001, p. 13.

p. 44    aliens "would have to know a good deal more": Stent, 1977, p. 132.

p. 45    "threaten to throw the very concept of 'the gene'": Keller, 2000, p. 67.

p. 45    "problem is not only that the music": Ibid., p. 63.

p. 45    the geneticist T. H. Morgan wrote: Quoted in Keller, 2000, p. 56.

## CHAPTER 4: BOUNDARY ISSUES

p. 51    "the heritability of male sexual orientation": Hamer & Copeland, 1998, p. 188.

p. 52    "[I]f two men lay bricks to build a wall": Levins & Lewontin, 1985, p. 111.

p. 52    "it is meaningless to ask whether iron rusts": Blackburn, 2002, p. 28.

p. 53    "Two diametrically opposed views of egg function": Cohen, 1979, pp. 5-6.

p. 54    gravity literally grounds the process of normal development: Eyal-Giladi, 1997.

p. 54    three primary tissue layers: see Purves & Lichtman, 1985.

p. 54    For the study of embryogenesis and gene expression, the fruit fly, Drosophila, has proven its worth: see Nüsslein-Volhard, 1996.

p. 55    once "we know how the system of causes works, assigning percentages to the different parts becomes an empty exercise": Wahlsten, 2002, p. 261.

p. 56    Michel Morange has wondered: Morange, 2001.

p. 56    Recent work shows that there is even more to be learned from PKU: Wahlsten, 2002, p. 260.

p. 57    "many animals, including all crocodilians studied thus far and some turtle and lizard species, completely lack sex chromosomes": Crews, 1994; Deeming and Ferguson, 1991.

p. 58    We notice a remarkable difference between these two siblings: see Vandenbergh, 2003.

p. 59    The anogenital distance in males is greater than that in females: Ibid.; vom Saal & Bronson, 1980.

p. 60    2M females produce litters with more males than females: Clark & Galef, 1995.

p. 61    a process called genetic imprinting: see Murphy & Jirtle, 2003.

p. 62    This was shown recently in a strain of mice: Waterland & Jirtle, 2003.

p. 62     the first cloned cat: Shin et al., 2002.

p. 63     Developmental psychobiology: for a recent survey of the field, see Michel & Moore, 1995.

p. 64     the "important question is not 'Is the animal isolated?' but 'From what is the animal isolated?'": Lehrman. 1953, p. 343.

p. 65     "adds nothing": Ibid., p. 344.

p. 66     "These traditions dictated that for Lorenz": Rosenblatt, 1995, pp. 238-239.

p. 67     Mother rats not only lick their pups, they ingest what the pups release: see Gubernick & Alberts, 1985.

p. 67     male pups are licked more often than female pups: for review, see Moore, 1995.

p. 70     the traits of the young reflected the traits of their foster, not their biological, mothers: for review, see Meaney, 2001.

p. 70     a stunning example of non-genetic transmission of behavioral traits: Francis et al., 1999.

p. 70     Meaney and his colleagues have taken their findings further: Liu et al., 1997, 2000.

p. 71     "A recent startling twist in this story": Weaver et al., 2004.

p. 72     "genes contain the information for the circuit diagram": Benzer, 1971, p. 1015.

p. 72     "an enormous proliferation of neurons": see Purves, 1994.

p. 73     "neural activity modulates the growth of nerve cells and neural circuits": Purves, 1994, p. xii.

p. 74     "the viewpoint that the structure and function of the nervous system of an animal is specified by its genes provides too narrow a context": Stent, 1981, p. 186.

p. 75     groundbreaking experiments beginning in the 1960s: for a retrospective on their work, see Hubel & Wiesel, 1998.

p. 76     neurons within the retina are spontaneously active: Galli & Maffei, 1988.

p. 76     spontaneous activity of retinal neurons can substitute for light in the development of ocular dominance columns: for review, see Shatz, 1990.

p. 77     experimenters surgically manipulated an embryonic frog: Constantine-Paton & Law, 1978.

p. 78     As an example of induction: see Guillery 1974, 1986.

p. 79     a mutant rat that exhibits selective degeneration of photoreceptors: Mullen & LaVail, 1976.

## CHAPTER 5: DEVELOPING AN INSTINCT

p. 81     William Paley and Erasmus Darwin's perspectives on behavior and cognition: see Richards, 1987.

p. 82     "that questions about what is innate and what is learned are as meaningful as our ancestors thought they were": Spelke & Newport, 1998, p. 276.

p. 82    "precursors that are rationally related to the behaviors for which they are supposedly responsible": Johnston, 1997, p. 511.

p. 82    "aims to describe the brute facts of natural phenomena and then tries to explain them": Ibid., p. 510.

p. 83    "Charles Darwin, in The Origin of Species, defined instinctive behavior": Darwin, 1859/1983, pp. 317-318.

p. 83    "form a natural unit of heredity": Lorenz, 1958, p. 121.

p. 83    "preoccupation with external influences on behavior": Ibid., p. 120.

p. 83    "most birds (as well as virtually all mammals and reptiles) scratch": Ibid., pp. 119-120.

p. 84    "part of their genetic heritage": Ibid., p. 119.

p. 85    "No bit of behavior": Beach, 1955, p. 407.

p. 85    "the concept of instinct will disappear": Ibid., p. 409.

p. 85    "braking an automobile and swinging a baseball bat": Hailman, 1969, p. 241.

p. 86    "The parent lowers its head and points its beak downward": Ibid.

p. 88    "the newly hatched chick responds best to a very simple stimulus situation": Ibid., p. 246.

p. 88    "the characteristics of the parent match the chick's ideal more closely than any other object in the environment": Ibid.

p. 89    "strongly suggests that the normal development of other instincts entails a component of learning": Ibid., p. 249.

p. 90    "This little creature": Pavlov, 1927, p. 9.

p. 90    "we are puzzled to find any line of demarcation": Ibid.

p. 91    "it has been used from the very beginning with a strictly scientific connotation": Ibid., p. 11.

p. 93    To study these phenomena in the laboratory: see Bolhuis & Honey, 1998; ten Cate, 1994.

p. 95    "This complexity will not exactly cover what most people have in mind": ten Cate, 1994, p. 125.

p. 95    "hatchlings were capable of identifying the maternal assembly call": Gottlieb, 1997, p. 29.

p. 96    Lorenz "blithely reminded all concerned": Ibid., p. 29.

p. 97    Gottlieb's discovery of an inborn preference of ducklings for the maternal assembly call: for review, see Gottlieb, 1997.

p. 101   "The search for nonobvious experiential bases to instinctive behavior": Ibid., pp. 76-77.

p. 102   "fatal to the whole theory": Darwin, 1859/1983, p. 352.

p. 102   "The great difficulty": Ibid.

p. 103   an "overriding principle": Hölldobler & Wilson, 1990, p. 348.

p. 104   "Bekoff has proposed": Bekoff, 1992, p. 1501.

p. 104   a striking preference to affiliate with each other: for review, see Alberts & Cramer, 1988.

p. 105   "rats exhibit a species-typical preference because they are reared in a species-typical environment": Ibid., p. 26.

p. 105   The maintenance of a regulated and stable thermal environment is critical for normal development: for a general introduction to thermoregulatory concepts, see Blumberg, 2001.

p. 106   even small, brief, shifts in temperature can have disastrous consequences for fetuses: Germain et al., 1985.

p. 106   pregnant rats were launched on the space shuttle Atlantis: Ronca & Alberts, 2000.

p. 107   rapid eye movements are produced by twitches of the eye muscles: see Chase & Morales, 1990.

p. 107   spontaneous movements contribute to the development of cartilage, tendons, ligaments, and bone: for review, see Müller, 2003.

p. 108   a surprising array of complex behavioral patterns: for review, see Robinson & Kleven, in press.

p. 109   "In few organisms": Ibid.


CHAPTER 6: OF HUMAN BONDAGE

p. 111   "ineffectual wastes of words": James, 1890/1983, p. 1012.

p. 111   "has a far greater variety of impulses than any lower animal": Ibid., p. 1010.

p. 112   "find the list too large, others too small": Ibid., p. 1056.

p. 112   "no other mammal, not even the monkey, shows so large an array": Ibid.

p. 112   "The human mind has certain innate or inherited tendencies": McDougall, 1960, p. 2.

p. 112   used "so loosely that they have almost spoilt them for scientific purposes": Ibid., p. 18.

p. 112   "a cloak of ignorance": Ibid., p. 19.

p. 113   "may be admitted to be a useful procedure": Dunlap, 1919, p. 309.

p. 113   "cease talking of 'instincts'": Ibid., p. 311.

p. 113   "there was nothing in our background in behaviorism": Breland & Breland, 1961, p. 289.

p. 113   "trapped by strong instinctive behaviors": Ibid., p. 290.

p. 113   when "behaviorism tossed out instinct": Ibid.

p. 114   as the fruits of behaviorism continued to fuel some of the most significant advances in neuroscience: see Thompson, 1994.

p. 115   "has a cognitive program that defines a red dot on the end of a beak as salient information from the environment": Tooby & Cosmides, 1992, p. 65.

p. 115   "what is special about the human mind": Ibid., p. 113.

p. 116  "set of computational machines," but each of these machines "was designed by natural selection to solve adaptive problems faced by our hunter-gatherer ancestors": Duchaine et al., 2001, p. 225.

p. 116  "By adding together a face recognition module": Tooby & Cosmides, 1992, p. 113.

p. 117  the nervous system does exhibit modularity: see Shettleworth, 2000.

p. 117  in the extreme case of the platypus: see Krubitzer, 1995.

p. 117  "there is no logically necessary connection between innateness and modularity": Shettleworth, 2000, p. 48.

p. 118  "The degree to which a particular aspect of information processing": Ibid.

p. 118  mammals possess two distinct thirst systems: for review, see Johnson & Thunhorst, 1997.

p. 121  "accidentally submerged in water": Craig, 1912, p. 275.

p. 122  Ted Hall and his students examined these issues: Hall et al., 2000; Changizi et al., 2002.

p. 123  "the normal arrangement of the animal's world": Hall et al., 2000, p. 105.

p. 123  "finding that a specific desire as rudimentary as that for water": Ibid.

p. 126  "conspicuously lacking in astuteness": Berger et al., 2001, p. 1038.

p. 126  "naïve prey have the capacity to process information about predators swiftly," even "in a single generation": Ibid., p. 1039.

p. 127  "fear and respect for reptiles is a likely core mammalian heritage": Öhman & Mineka, 2003, p. 5.

p. 127  raised on live insects: Masataka, 1993.

p. 127  "more advanced human cognition": Öhman & Mineka, 2003, p. 7.

p. 127  "This specialized behavioral module": Ibid.

p. 128  "Genetic variability might explain why not all individuals show fear of snakes": Ibid., p. 6.

p. 128  snakes evolved from a non-dinosaur ancestor at least thirty-five million years before the extinction of the dinosaurs: see Rage, 1987.

p. 130  Such an apparent predisposition was forcefully demonstrated thirty years ago: for review of this and subsequent findings, see ten Cate, 1994.

p. 132  "because nothing can be inserted in it through experience": Mayr, 1976, p. 696.

p. 132  "There are several groups of birds": Ibid., p. 697.

p. 134  "an independent system designed to insure identification": King & West, 1977, p. 1004.

p. 134  the "'just-so-ish' gist of the story": West & King, 2001, p. 596.

p. 135  "the safety net is not prebuilt into either sex, but socially built up between them": Ibid., p. 606.

p. 135  like "teenagers at a dance where the boys talk together across the room from the girls": Ibid., p. 598.

p. 135  "sweeping dismissals of our original proposition": Ibid.

p. 137    "a pitcher, fielders, catchers, umpires, and the performance of the previous bat-
          ters": Ibid., p. 610.

p. 137    "the illusory view": Ibid.

p. 137    "contained completely in the original fertilized zygote": Mayr, 1976, p. 697.

p. 139    When amniotic fluid and breast milk become suffused with flavors: see Mennella
          et al., 2001.

p. 140    young rats learn about food from adults through social transmission: for review,
          see Galef, 2003.

p. 141    "black rat pups acquire the pine cone stripping technique": Terkel, 1995, p. 149.

p. 142    "the key factor in the pups' learning process": Ibid.

p. 142    "a better understanding of the origins of behavioral traditions": Galef, 2003,
          p. 177.

p. 142    "the human psychological architecture": Tooby & Cosmides, 1992, p. 34.

p. 143    list of human instincts: Pinker, 1994, p. 420.

p. 143    "But if there is a language instinct": Pinker, 1994, p. 299.

p. 144    Chomsky countered: Chomsky, 1959.

p. 145    "I think Chomsky and his followers have articulated a central conundrum about
          language learning": Deacon, 1997, p. 105.

p. 146    "Children's minds need not innately embody language structures": Ibid., p. 109.

p. 147    a "neurological bias acting as a relentless force in social evolution": Ibid., p. 119.

p. 147    "does not necessarily imply another even though people often assume, without
          evidence, that it does": Bateson, 2002, p. 2212.

## CHAPTER 7: THE NATIVISTS ARE RESTLESS

p. 150    "modules, or families of instincts": Pinker, 1994, p. 299.

p. 150    "In the hunter-gatherer societies": Spelke & Newport, 1998, p. 314.

p. 151    the story of Clever Hans: see Candland, 1993.

p. 155    "for individuals with severe disabilities": see http://soeweb.syr.edu/thefci/

p. 155    "can see the letters in his mind and can aim his fingers at an imaginary key-
          board": From television broadcast of "Free from silence," PrimeTime Live, Jan.
          23, 1992.

p. 158    "the first study of the development of geometrical knowledge": Spelke & New-
          port, 1998, p. 311.

p. 158    The exchange with the boy: Plato, 1966, pp. 108-114.

p. 159    a cascade of rapid-fire conclusions: Ibid., pp. 114-115.

p. 159    "virtue will be acquired neither by nature nor by teaching": Ibid., p. 128.

p. 159    "the direction that the inquiry takes": Klein, 1965, p. 103.

p. 159    "it is Socrates who draws all the figures": Ibid.

p. 160    "leading questions which any judge would disallow": Russell, 1945, p. 93.

p. 160    "it is quite unavailing when the object is to discover new facts": Ibid.

p. 161   Jean Piaget had asserted many years before: Piaget, 1951.

p. 162   "holds the key to our understanding": Meltzoff & Decety, 2003, p. 491.

p. 162   "Metaphorically, we can say that nature endows humans": Ibid.

p. 163   "replicated and extended": Ibid., p. 492.

p. 163   "there is little basis for the hypothesis that neonates can imitate oral gestures": Anisfeld et al., 2001, p. 121.

p. 166   "cannot execute the full sequence of behaviors that will bring the object to the mouth": Jones, 1996, p. 1966.

p. 166   "puts the final component of autonomous oral exploration into place": Ibid.

p. 166   "to grasp and transport objects to their mouths": Ibid.

p. 169   "exists a mental mechanism, dedicated to representing reasoning about number": Wynn, 1998, p. 297.

p. 169   "appreciate the precise numerical relationships that hold between small numbers": Ibid.

p. 170   "possess a sensitivity to number per se": Ibid., p. 296.

p. 170   "infants attend to contour length, rather than number, to discriminate between sets": Clearfield & Mix, 1999, p. 410.

p. 171   "because contour length is correlated": Ibid.

p. 171   "infants prefer to discriminate on the basis of basic perceptual variables": Ibid.

p. 171   when Steven Pinker inventoried human instincts: Pinker, 1994, p. 420.

p. 172   "appreciate the precise numerical relationships that hold between small numbers": Wynn, 1998, p. 297.

p. 172   "toe-hold upon which to enter the realm of mathematical thought": Ibid., p. 302.

p. 174   "Renée Baillargeon also performs a test of solidity but uses a 'drawbridge'": For examples of critical analyses of this single experimental paradigm, see Schilling, 2000; Bogartz et al., 2000; Cashon & Cohen, 2000.

p. 174   "seemingly very small changes in stimuli": Schöner & Thelen, in press.

p. 175   "may invent experiments to create 'violations of expectancy' but infants experience them as visual events": Ibid.

p. 176   Esther Thelen performed groundbreaking work: Thelen, 1982.

p. 176   "present before it is needed for that purpose": Spelke & Newport, 1998, p. 281.

p. 176   a "competence that does not normally express itself in behavior": Ibid.

p. 178   a simple hide-and-seek game first described by Jean Piaget in the 1950s: Piaget, 1954.

p. 178   "[A]n infant sits before two hiding locations": Smith et al., 1999, p. 236.

p. 179   "If the A-not-B error is a true measure": Thelen et al., 2001, p. 3.

p. 180   "is not about what infants have and don't have as enduring concepts": Ibid., p. 4.

p. 181   This was accomplished in children at two and three years of age: Spencer et al., 2001.

p. 182    "looking at or away from an event display is a motor act": Thelen et al., 2001, p. 33.

p. 183    infants "are capable of reasoning": Spelke et al., 1992, p. 606.

p. 183    "come to know about states of the world that they never perceived": Ibid.

p. 183    "I challenge anyone": Baillargeon, 1999a, p. 157.

p. 184    "development leads to the enrichment of conceptions around an unchanging core": Spelke et al., 1992, p. 605.

p. 184    "Intuitive mechanics: knowledge of the motions, forces, and deformations that objects undergo": Pinker, 1994, p. 420.

p. 186    "[I]n the first weeks of walking": Adolph, 2000, p. 291.

p. 187    "may be posture-specific because each postural milestone represents a different perception-action system with different control variables": Ibid.

p. 187    "different vantage points for viewing the ground": Ibid.

p. 187    "If infants learn to avoid a drop-off because they are afraid of heights": Adolph, 2002, p. 25.

p. 188    "learning about balance control may occur over many thousands of trials": Ibid., p. 32.

p. 188    "Danger, including the emotions of fear and caution": Pinker, 1994, p. 420.

p. 188    "phobias for stimuli such as heights": Ibid.

p. 190    "there was not a single instance": Simcock & Hayne, 2002, p. 229.

p. 192    "perceive a world of complete and solid objects, not visible fragments": Kellman & Spelke, 1983, p. 483.

p. 193    "roots perception of partly hidden objects in an unlearned conception of the physical world": Ibid., p. 521.

p. 193    "may be an essential property, at the center of our conception of the world": Ibid., p. 522.

p. 193    "the visible evidence literally": Slater et al., 1990, p. 48.

p. 193    "that infants from birth have an innate idea of the underlying unity and coherence of objects": Ibid.

p. 195    "consistent with the idea that infant attention to human stimuli may be biased toward the gender with which the infant has greater experience": Quinn, in press.

p. 197    "[I]n order to explain the first steps in the development of specialization for face processing": Turati, 2004, p. 8.

p. 197    "impair specialized areas of processing": Karmiloff-Smith, 2000, p. 177.

p. 198    a relatively open face-recognition capability is narrowed during the first postnatal year as human infants gain experience with human faces: Pascalis et al., 2002.

p. 198    this same principle applies to language production: Werker & Vouloumanos, 2001.

p. 199    "the abnormal brain is not a normal brain with parts intact and parts impaired": Karmiloff-Smith, 2000, p. 178.

CHAPTER 8: STRAYING FROM THE HERD

p. 204 "responsive phenotype at the center of development" for which it "is of little developmental consequence": West-Eberhard, 2003, p. 99.

p. 204 "death knell of the nature-nurture controversy": Ibid.

p. 204 "Confused concepts are virtually inevitable if data on real mechanisms are ignored": Ibid., p. 11.

p. 206 "A flock of birds sweeps across the sky": Resnick, 1994, p. 3.

p. 206 "follows a set of simple rules, reacting to the movements of the birds nearby": Ibid.

p. 206 "transcription of DNA that underlies the involvement of the genes": Johnston & Edwards, 2002, p. 27.

p. 207 "any influence that acts through the developing animal's sense organs and is processed by its nervous system": Ibid., p. 26.

p. 207 "all other environmental effects": Ibid.

p. 208 "Domestic instincts are sometimes spoken of as actions which have become inherited solely from long-continued and compulsory habit": Darwin, 1859/1983, p. 325.

p. 208 "It may be doubted whether any one would have thought of training a dog to point": Ibid.

p. 209 "the right stuff to start with": Adams & Meisner, 2002.

p. 209 "This tape will not teach you how to train your dog to point because that's not possible": Ibid.

p. 209 "only the exaggerated pause of an animal preparing to spring on its prey": Darwin, 1859/1983, p. 325.

p. 210 numerous "genetic" disorders have been identified: see Ostrander & Kruglyak, 2000.

p. 211 in a flurry of scientific advances over the last seven years: see Lin et al., 1999; Taheri et al., 2002.

p. 211 "the result of physical characteristics associated with the breed standard": Ostrander & Kruglyak, 2000, p. 1273.

p. 212 "Other studies have turned up some remarkably narrow and distinctive behavioral lineages": Budiansky, 1999, p. 46.

p. 212 a newborn mouse can exhibit grooming-like movements with its forepaws: see Fentress & McLeod, 1986.

p. 215 breeders have found that superior pointing dogs do not give rise to superior pointing progeny: Willis, 1995, p. 54.

p. 215 expression by offspring of those traits considered most important in a useful hunting dog are predicted best by the identity of the mother: Ibid., pp. 54-55.

p. 217 the most effective guarding dogs are those that have, ideally before the age of eight weeks, been reared with livestock: Ibid, pp. 56-57.

p. 217    Zing-Yang Kuo reported in 1930: see Kuo, 1967.

p. 218    Fleming chose to compare Border collies with Newfoundlands: Fleming, personal communication; Fleming's experiment reported previously by Budiansky, 1999, p. 46.

p. 220    "are eager to establish human contact, whimpering to attract attention and sniffing and licking experimenters like dogs": Trut, 1999, p. 163.

p. 220    "When a pup is one month old, an experimenter offers it food from his hand while trying to stroke and handle the pup": Ibid.

p. 221    changes in the timing of development, or heterochrony: for detailed discussions of this issue, see Gould, 1977; McKinney & McNamara, 1991.

# SOURCES AND
# SUGGESTED READING

Adams, H., and D. Meisner. "Point! Refining the Pointing Instinct." Minocqua, WI: Willow Creek Press, 2002.

Adolph, K. E. "Learning to Keep Balance." In *Advances in Child Development & Behavior,* edited by R. Kail, 1–40. Amsterdam: Elsevier, 2002.

———. "Specificity of Learning: Why Infants Fall over a Veritable Cliff." *Psychological Science* 11 (2000): 290–95.

Aisner, R., and J. Terkel. "Ontogeny of Pine Cone Opening Behaviour in the Black Rats, *Rattus Rattus." Animal Behaviour* 44 (1992): 327–36.

Alberts, J. R., and C. P. Cramer. "Ecology and Experience: Sources of Means and Meaning of Developmental Change." In *Handbook of Behavioral Neurobiology,* edited by E. M. Blass, 1–39. New York: Plenum Press, 1988.

Anisfeld, M. "Only Tongue Protrusion Modeling Is Matched by Neonates." *Developmental Review* 16 (1996): 149–61.

Anisfeld, M., G. Turkewitz, S. A. Rose, F. R. Rosenberg, F. J. Sheiber, D. A. Couturier-Fagan, J. S. Ger, and I. Sommer. "No Compelling Evidence That Newborns Imitate Oral Gestures." *Infancy* 2 (2001): 111–22.

Aronson, L. R., E. Tobach, J. S. Rosenblatt, and D. S. Lehrman, eds. *Selected Writings of T. C. Schneirla.* San Francisco: W. H. Freeman and Company, 1972.

Baillargeon, R. "Response to Smith and the Commentators." *Developmental Science* 2 (1999*a*): 157–61.

———. "Young Infants' Expectations About Hidden Objects: A Reply to Three Challenges." *Developmental Science* 2 (1999*b*): 115–32.

Baltimore, D. "The Brain of a Cell." *Science* 84 (1984): 150.

Bateson, P. "The Active Role of Behaviour in Evolution." In *Evolutionary Processes and Metaphors,* edited by M-W. Ho and S. W. Fox, 191–207. Chichester: John Wiley & Sons Ltd., 1988.

———. "Are There Principles of Behavioural Development?" In *The Development and Integration of Behaviour,* edited by P. Bateson. Cambridge: Cambridge University Press, 1991.

———. "The Corpse of a Wearisome Debate." *Science* 297 (2002): 2212–13.

———. "Taking the Stink out of Instinct." In *Alas, Poor Darwin: Arguments against Evolutionary Psychology,* edited by H. Rose and S. Rose, 189–207. New York: Harmony Books, 2000.

Bateson, P., and P. Martin. *Design for a Life: How Behavior and Personality Develop.* New York: Simon & Schuster, 2000.

Beach, F. A. "The Descent of Instinct." *Psychological Review* 62 (1955): 401–10.

Behe, M. J. *Darwin's Black Box: The Biochemical Challenge to Evolution.* New York: The Free Press, 1996.

Bekoff, A. "Neuroethological Approaches to the Study of Motor Development in Chicks: Achievements and Challenges." *Journal of Neurobiology* 23 (1992): 1486–505.

Benzer, S. "From the Gene to Behavior." *Journal of the American Medical Association* 218 (1971): 1015–22.

Berger, J., J. E. Swenson, and I. Persson. "Recolonizing Carnivores and Naive Prey: Conservation Lessons from Pleistocene Extinctions." *Science* 291 (2001): 1036–39.

Biklen, D. "Communication Unbound: Autism and Praxis." *Harvard Educational Review* 60 (1990): 291–314.

Blackburn, S. "Meet the Flintstones." *The New Republic,* November 25 2002, 28–33.

Blumberg, M. S. *Body Heat: Temperature and Life on Earth.* Cambridge, MA: Harvard University Press, 2002.

Blumberg, M. S., and E. A. Wasserman. "Animal Mind and the Argument from Design." *American Psychologist* 50 (1995): 133–44.

Bogartz, R. S., J. L. Shinskey, and T. H. Schilling. "Object Permanence in Five-and-a-Half-Month-Old Infants?" *Infancy* 1 (2000): 403–28.

Bolhuis, J. J., and R. C. Honey. "Imprinting, Learning and Development: From Behaviour to Brain and Back." *Trends in Neurosciences* 21 (1998): 306–11.

Bouchard, T. J. "Genes, Environment, and Personality." Science 264 (1994): 1700–01.

Breland, K., and M. Breland. "The Misbehavior of Organisms." *American Psychologist* 16 (1961): 681–84.

Brenner, S. "The Genetics of Behaviour." *British Medical Bulletin* 29 (1973): 269–71.

Budiansky, S. "The Truth About Dogs." *Atlantic Monthly,* July 1999, 39–53.

Burtt, E. H., Jr., J. Bitterbaum, and J. P. Hailman. "Head-Scratching Method in Swallows Depends on Behavioral Context." *Wilson Bulletin* 100 (1988): 679–82.

Burtt, E. H., Jr., and J. P. Hailman. "Head-Scratching among North American Wood-Warblers (*Parulidae*)." *Ibis* 120 (1978): 153–70.

Candland, D. K. *Feral Children & Clever Animals.* Oxford: Oxford University Press, 1993.

Cashon, C. H., and L. B. Cohen. "Eight-Month-Old Infants' Perception of Possible and Impossible Events." *Infancy* 1 (2000): 429–46.

Changizi, M. A., R. M. F. McGehee, and W. G. Hall. "Evidence That Appetitive Responses for Dehydration and Food-Deprivation Are Learned." *Physiology & Behavior* 75 (2002): 295–304.

Chase, M. H., and F. R. Morales. "The Atonia and Myoclonia of Active (REM) Sleep." *Annual Review of Psychology* 41 (1990): 557–84.

Chomsky, N. "A Review of B. F. Skinner's "Verbal Behavior"." *Language* 35 (1959): 26–58.

Clark, M. M., and B. G. Galef, Jr. "A Gerbil Dam's Fetal Intrauterine Position Affects the Sex Ratios of Litters She Gestates." *Physiology & Behavior* 57 (1995): 297–99.

Clayton, N. S. "The Influence of Social Interactions on the Development of Song and Sexual Preferences in Birds." In *Causal Mechanisms of Behavioural Development,* edited by J. A. Hogan and J. J. Bolhuis, 98–115. Cambridge: Cambridge University Press, 1994.

Clearfield, M. W., and K. S. Mix. "Number Versus Contour Length in Infants' Discrimination of Small Visual Sets." *Psychological Science* 10 (1999): 408–11.

Clutton-Brock, J. "Origins of the Dog: Domestication and Early History." In *The Domestic Dog: Its Evolution, Behaviour, and Interactions with People,* edited by J. Serpell, 22–47. Cambridge: Cambridge University Press, 1995.

Cohen, J. "Maternal Constraints on Development." In *Maternal Effects in Development,* edited by D. R. Newth and M. Balls, 1–28. Cambridge: Cambridge University Press, 1979.

Collins, F. S., L. Weiss, and K. Hudson. "Heredity and Humanity." *The New Republic,* June 25 2001, 27–29.

Constantine-Paton, M., and M. I. Law. "Eye-Specific Termination Bands in Tecta of Three-Eyed Frogs." *Science* 202 (1978): 639–41.

Coppinger, R., and R. Schneider. "Evolution of Working Dogs." In *The Domestic Dog: Its Evolution, Behaviour, and Interactions with People,* edited by J. Serpell, 22–47. Cambridge: Cambridge University Press, 1995.

Craig, W. "Appetites and Aversions as Constituents of Instincts." *Biological Bulletin* 34 (1918): 91–107.

———. "Observations on Doves Learning to Drink." *Journal of Animal Behavior* 2 (1912): 273–79.

Crews, D. "Animal Sexuality." *Scientific American* 270 (1994): 108–14.

Darwin, C. *The Origin of Species.* Harmondsworth: Penguin Books Ltd, 1859/1983.

Dawkins, R. *The Blind Watchmaker.* New York: W. W. Norton, 1986.

————. *The Selfish Gene.* Second ed. Oxford: Oxford University Press, 1989.

Deacon, T. W. "Multilevel Selection in a Complex Adaptive System: The Problem of Language Origins." In *Evolution and Learning: The Baldwin Effect Reconsidered,* edited by B. H. Weber and D. J. Depew, 81–106. Cambridge, MA: MIT Press, 2003.

————. *The Symbolic Species.* New York: W. W. Norton, 1997.

DeCasper, A. J., and W. P. Fifer. "Of Human Bonding: Newborns Prefer Their Mothers' Voices." *Science* 208 (1980): 1174–76.

Deeming, D. C., and M. W. J. Ferguson, eds. *Egg Incubation: Its Effects on Embryonic Development in Birds and Reptiles.* Cambridge: Cambridge University Press, 1991.

Diamond, S. "Four Hundred Years of Instinct Controversy." *Behavior Genetics* 4 (1974): 237–52.

————. "Gestation of the Instinct Concept." In *Historical Conceptions of Psychology,* edited by M. Henle, J. Jaynes and J. J. Sullivan, 150–65. New York: Springer Publishing Company, Inc., 1973.

Duchaine, B., L. Cosmides, and J. Tooby. "Evolutionary Psychology and the Brain." *Current Opinion in Neurobiology* 11 (2001): 225–30.

Dunlap, K. "Are There Any Instincts?" *Journal of Abnormal Psychology* 14 (1919–1920): 307–11.

Elman, J. L., E. A. Bates, M. H. Johnson, A. Karmiloff-Smith, D. Parisi, and K. Plunkett. *Rethinking Innateness: A Connectionist Perspective on Development.* Cambridge: The MIT Press, 1996.

Eyal-Giladi, H. "Establishment of the Axis in Chordates: Facts and Speculations." *Development* 124 (1997): 2285–96.

Fentress, J. C. "History of Developmental Neuroethology: Early Contributions from Ethology." *Journal of Neurobiology* 23 (1992): 1355–69.

Fentress, J. C., and P. J. McLeod. "Motor Patterns in Development." In *Handbook of Behavioral Neurobiology,* edited by E. M. Blass, 35–97. New York: Plenum Press, 1986.

Fleming, A. S., G. W. Kraemer, A. Gonzalez, V. Lovic, S. Rees, and A. Melo. "Mothering Begets Mothering: The Transmission of Behaviour and Its Neurobiology across Generations." *Pharmacology, Biochemistry & Behavior* 73 (2002): 61–75.

Francis, D., J. Diorio, D. Liu, and M. J. Meaney. "Nongenomic Transmission across Generations of Maternal Behavior and Stress Responses in the Rat." *Science* 286 (1999): 1155–8.

Galef, B. G., Jr. "Traditional Behaviors of Brown and Black Rats." In *The Biology of Traditions: Models and Evidence,* edited by S. Perry and D. Fragaszy, 159–86. Chicago: University of Chicago Press, 2003.

Galli, L., and L. Maffei. "Spontaneous Impulse Activity of Rat Retinal Ganglion Cells in Prenatal Life." *Science* 242 (1988): 90–91.

Germain, M-A., W. S. Webster, and M. J. Edwards. "Hyperthermia as a Teratogen:

Parameters Determining Hyperthermia-Induced Head Defects in the Rat." *Teratology* 31 (1985): 265–72.

Gootman, E. "Separated at Birth in Mexico." *New York Times,* March 3 2003.

Gottlieb, G. *Individual Development and Evolution.* New York: Oxford University Press, 1992.

———. *Synthesizing Nature-Nurture: Prenatal Roots of Instinctive Behavior.* Mahwah: Lawrence Erlbaum Associates, 1997.

Gould, J. L., and C. G. Gould. *The Animal Mind.* New York: Scientific American Library, 1999.

Gould, S. J. *Ontogeny and Phylogeny.* Cambridge: The Belknap Press of Harvard University Press, 1977.

———. *The Panda's Thumb.* New York: W. W. Norton & Company, 1982.

Green, G. "Facilitated Communication: Mental Miracle or Sleight of Hand?" *Skeptic* 2 (1994): 68–76.

Gubernick, D. J., and J. R. Alberts. "Maternal Licking by Virgin and Lactating Rats: Water Transfer from Pups." *Physiology & Behavior* 34 (1985): 501–6.

Guillery, R. W. "Neural Abnormalities of Albinos." Trends in Neurosciences (1986): 364–67.

———. "Visual Pathways in Albinos." *Scientific American* 230 (1974): 44–54.

Hailman, J. P. "How an Instinct Is Learned." *Scientific American* (1969).

Haith, M. M. "Who Put the Cog in Infant Cognition? Is Rich Interpretation Too Costly." *Infant Behavior and Development* 21 (1998): 167–79.

Hall, W. G., H. M. Arnold, and K. P. Myers. "The Acquisition of an Appetite." *Psychological Science* 11 (2000): 101–05.

Hall, W. G., and R. W. Oppenheim. "Developmental Psychobiology: Prenatal, Perinatal, and Early Postnatal Aspects of Behavioral Development." *Annual Review of Psychology* 38 (1987): 91–128.

Hamer, D., and P. Copeland. *Living with Our Genes.* New York: Anchor Books, 1998.

Hauser, M. D. *Wild Minds: What Animals Really Think.* New York: Henry Holt, 2000.

Hinde, R. A. "Dichotomies in the Study of Development." In *Genetic and Environmental Influences on Behaviour,* edited by J. M. Thoday and A. S. Parkes, 3–14. Edinburgh: Oliver & Boyd, 1968.

Holland, V. S. *Herding Dogs: Progressive Training.* New York: Howell Book House, 1994.

Holmes, J. "Understanding the Border Collie." In *The Ultimate Border Collie,* edited by A. Hornsby. New York: Simon and Schuster, 1998.

Horn, G. "Visual Imprinting and the Neural Mechanisms of Recognition Memory." *Trends in Neurosciences* 21 (1998): 300–05.

Hubbard, R., and E. Wald. *Exploding the Gene Myth.* Boston: Beacon Press, 1993.

Hubel, D. D., and T. N. Wiesel. "Early Exploration of the Visual Cortex." *Neuron* 20 (1998): 401–12.

Hume, D. "Dialogues Concerning Natural Religion." In *Dialogues Concerning Natural*

*Religion and the Posthumous Essays,* edited by R. H. Popkin, 1–89. Indianapolis: Hackett, 1776/1985.

Hölldobler, B., and E. O. Wilson. *The Ants.* Cambridge, MA: Harvard University Press, 1990.

Inglis, F. M., K. E. Zuckerman, and R. G. Kalb. "Experience-Dependent Development of Spinal Motor Neurons." *Neuron* 26 (2000): 299–305.

Jablonka, E., and M. J. Lamb. *Epigenetic Inheritance and Evolution: The Lamarckian Dimension.* Oxford: Oxford University Press, 1995.

Jablonka, E., M. J. Lamb, and E. Avital. "'Lamarckian' Mechanisms in Darwinian Evolution." *Trends in Ecology and Evolution* 13 (1998): 206–10.

James, W. *The Principles of Psychology.* Cambridge: Cambridge University Press, 1890/1983.

Johnson, A. K., and R. L. Thunhorst. "The Neuroendocrinology of Thirst and Salt Appetite: Visceral Sensory Signals and Mechanisms of Central Integration." *Frontiers in Neuroendocrinology* 18 (1997): 292–353.

Johnston, T. D. "Comment on Miller." In *Evolving Explanations of Development: Ecological Aapproaches to Organism-Environment Systems,* edited by C. Dent-Read and P. Zukow-Goldring, 509–13. Washington, DC: American Psychological Association, 1997.

———. "The Persistence of Dichotomies in the Study of Behavioral Development." *Developmental Review* 7 (1987): 149–82.

Johnston, T. D., and L. Edwards. "Genes, Interactions, and the Development of Behavior." *Psychological Review* 109 (2002): 26–34.

Jones, S. S. "Imitation or Exploration? Young Infants' Matching of Adults' Oral Gestures." *Child Development* 67 (1996): 1952–69.

Joseph, J. *The Gene Illusion: Genetic Research in Psychiatry and Psychology under the Microscope.* Ross-on-Wye: PCCS Books, 2003.

———. "Separated Twins and the Genetics of Personality Differences: A Critique." *American Journal of Psychology* 114 (2001): 1–30.

Karmiloff-Smith, A. "Why Babies' Brain Are Not Swiss Army Knives." In *Alas, Poor Darwin: Arguments against Evolutionary Psychology,* edited by H. Rose and S. Rose, 173–87. New York: Harmony Books, 2000.

Kauffman, S. *At Home in the Universe: The Search for the Laws of Self-Organization and Complexity.* New York: Oxford University Press, 1995.

Keller, E. F. *The Century of the Gene.* Cambridge, MA: Harvard University Press, 2000.

Kellman, P. J., and E. S. Spelke. "Perception of Partly Occluded Objects in Infancy." *Cognitive Psychology* 15 (1983): 483–524.

King, A. P., and M. J. West. "Species Identification in the North American Cowbird: Appropriate Responses to Abnormal Song." *Science* 195 (1977): 1002–04.

Klein, J. *A Commentary on Plato's Meno.* Chicago: University of Chicago Press, 1965.

Krubitzer, L. "The Organization of Neocortex in Mammals: Are Species Differences Really Different?" *Trends in Neuroscience* 18 (1995): 408–17.

Kuo, Z.-Y. *The Dynamics of Behavior Development: An Epigenetic View.* New York: Random House, 1967.

Lander, E. S., et al. "Initial Sequencing and Analysis of the Human Genome." *Nature* 409 (2001): 860–921.

Lehrman, D. S. "A Critique of Konrad Lorenz's Theory of Instinctive Behavior." *The Quarterly Review of Biology* 4 (1953): 337–63.

Levins, R., and R. Lewontin. "The Analysis of Variance and the Analysis of Causes." In *The Dialectical Biologist.* Cambridge, MA: Harvard University Press, 1985.

Lewontin, R. *The Triple Helix: Gene, Organism, and Environment.* Cambridge, MA: Harvard University Press, 2000.

Lewontin, R. C. *Biology as Ideology.* New York: Harper Perennials, 1992.

Lewontin, R. C., S. Rose, and L. J. Kamin. *Not in Our Genes.* New York: Pantheon Books, 1984.

Lickliter, R., and H. Honeycutt. "Developmental Dynamics: Toward a Biologically Plausible Evolutionary Psychology." *Psychological Bulletin* 129 (2003): 819–35.

Lin, L., J. Faraco, R. Li, H. Kadotani, W. Rogers, X. Lin, X. Qiu, P. J. de Jong, S. Nishino, and E. Mignot. "The Sleep Disorder Canine Narcolepsy Is Caused by a Mutation in the Hypocretin (Orexin) Receptor 2 Gene." *Cell* 98 (1999): 365–76.

Liu, D., J. Diorio, J. C. Day, D. D. Francis, and M. J. Meaney. "Maternal Care, Hippocampal Synaptogenesis and Cognitive Development in Rats." *Nature Neuroscience* 3 (2000): 799–806.

Liu, D., J. Diorio, B. Tannenbaum, C. Caldji, D. Francis, A. Freedman, S. Sharma, D. Pearson, P. M. Plotsky, and M. J. Meaney. "Maternal Care, Hippocampal Glucocorticoid Receptors, and Hypothalamic-Pituitary-Adrenal Responses to Stress." *Science* 277 (1997): 1659–62.

Lorenz, K. Z. "The Evolution of Behavior." *Scientific American* (1958).

———. *The Foundations of Ethology.* New York: Springer-Verlag, 1981.

Macchi Cassia, V., C. Turati, and F. Simion. "Can a Nonspecific Bias toward Top-Heavy Patterns Explain Newborns' Face Preference?" *Psychological Science* 15 (2004): 379–83.

Martin, P., and P. Bateson. "Behavioural Development in the Cat." In *The Domestic Cat: The Biology of Its Behaviour,* edited by D. C. Turner and P. Bateson, 9–22. Cambridge: Cambridge University Press, 1988.

Masataka, N. "Effects of Experience with Live Insects on the Development of Fear of Snakes in Squirrel Monkeys, *Saimiri Sciureus.*" *Animal Behaviour* 46 (1993): 741–46.

Mayr, E. "Behavior Programs and Evolutionary Strategies." In *Evolution and the Diversity of Life,* 694–711. Cambridge, MA: Harvard University Press, 1976.

McDougall, W. *An Introduction to Social Psychology.* London: Methuen, 1960.

McKinney, M. L., and K. J. McNamara. *Heterochrony: The Evolution of Ontogeny.* New York: Plenum Press, 1991.

Meaney, M. J. "Maternal Care, Gene Expression, and the Transmission of Individual Differences in Stress Reactivity across Generations." *Annual Review of Neuroscience* 24 (2001): 1161–92.

Melton, D. A. "Pattern Formation During Animal Development." Science 252 (1991): 234–41.

Meltzoff, A. N., and J. Decety. "What Imitation Tells Us About Social Cognition: A Rapprochement between Developmental Psychology and Cognitive Neuroscience." *Philosophical Transactions of the Royal Society of London. Series B: Biological Sciences* 358 (2003): 491–500.

Meltzoff, A. N., and M. K. Moore. "Imitation of Facial and Manual Gestures by Human Neonates." *Science* 198 (1977): 75–78.

Mennella, J. A., C. P. Jagnow, and G. K. Beauchamp. "Prenatal and Postnatal Flavor Learning by Human Infants." *Pediatrics* 107 (2001): e88.

Michel, G. F., and C. L. Moore. *Developmental Psychobiology.* Cambridge: The MIT Press, 1995.

Miller, D. B. "The Effects of Nonobvious Forms of Experience on the Development of Instinctive Behavior." In *Evolving Explanations of Development: Ecological Aapproaches to Organism-Environment Systems,* edited by C. Dent-Read and P. Zukow-Goldring, 457–507. Washington, DC: American Psychological Association, 1997.

Mix, K. S. "The Construction of Number Concepts." *Cognitive Development* 17 (2002): 1345–63.

Mix, K. S., J. Huttenlocher, and S. C. Levine. "Multiple Cues for Quantification in Infancy: Is Number One of Them?" *Psychological Bulletin* 128 (2002): 278–94.

Moore, C. L. "Maternal Contributions to Mammalian Reproductive Development and the Divergence of Males and Females." *Advances in the Study of Behavior* 24 (1995): 47–118.

Moore, D. S. *The Dependent Gene: The Fallacy of "Nature vs. Nurture".* New York: W. H. Freeman & Company, 2001.

Morange, M. *The Misunderstood Gene.* Translated by M. Cobb. Cambridge, MA: Harvard University Press, 2001.

Morgan, C. L. *Habit and Instinct.* New York: Arno Press, 1896/1973.

Mullen, R. J., and M. M. LaVail. "Inherited Retinal Dystrophy: Primary Defect in Pigment Epithelium Determined with Experimental Rat Chimera." *Science* 192 (1976): 799–801.

Murphy, S. K., and R. L. Jirtle. "Imprinting Evolution and the Price of Silence." *Bioessays* 25 (2003): 577–88.

Müller, G. B. "Embryonic Motility: Environmental Influences and Evolutionary Innovation." *Evolution & Development* 5 (2003): 56–60.

Nijhout, H. F. "Metaphors and the Role of Genes in Development." *BioEssays* 12 (1990):
441–46.

Nüsslein-Volhard, C. "Gradients That Organize Embryo Development." *Scientific American* August (1996): 54–61.

Öhman, A., and S. Mineka. "The Malicious Serpent: Snakes as a Prototypical Stimulus for
an Evolved Module of Fear." *Current Directions in Psychological Science* 12 (2003): 5–9.

Oppenheim, R. W. "Pathways in the Emergence of Developmental Neuroethology:
Antecedents to Current Views of Neurobehavioral Ontogeny." *Journal of Neurobiology* 23 (1992): 1370–403.

Ostrander, E. A., and L. Kruglyak. "Unleashing the Canine Genome." *Genome Research*
101 (2000): 1271–74.

Oyama, S. *The Ontogeny of Information: Developmental Systems and Evolution.* Cambridge:
Cambridge University Press, 1985.

Oyama, S., P. E. Griffiths, and R. D. Gray, eds. *Cycles of Contingency: Developmental Systems and Evolution.* Cambridge: MIT Press, 2001.

Parker, H. G., L. V. Kim, N. B. Sutter, S. Carlson, T. D. Lorentzen, T. B. Malek, G. S.
Johnson, H. B. DeFrance, E. A. Ostrander, and L. Kruglyak. "Genetic Structure of
the Purebred Domestic Dog." *Science* 304 (2004): 1160–64.

Pascalis, O., M. de Haan, and C. A. Nelson. "Is Face Processing Species-Specific During
the First Year of Life?" *Science* 296 (2002): 1321–23.

Pavlov, I. P. *Conditioned Reflexes: An Investigation of the Physiological Activity of the Cerebral
Cortex.* Translated by G. V. Anrep. Humphrey Milford: Oxford University Press, 1927.

Petroski, H. *The Evolution of Useful Things.* New York: Vintage Books, 1992.

———. *To Engineer Is Human.* New York: Vintage Books, 1992.

Piaget, J. *The Construction of Reality in the Child.* New York: Basic Books, 1954.

———. *Play, Dreams, and Imitation in Childhood.* New York: Norton, 1951.

Pinker, S. *The Blank Slate: The Modern Denial of Human Nature.* New York: Viking, 2002.

———. *The Language Instinct: How the Mind Creates Language.* New York: HarperPeren-
nial, 1995.

Plato. "Meno." In *Greek Philosophy: Thales to Aristotle,* edited by R. E. Allen, 97–128. New
York: Free Press, 1966.

Purves, D. *Neural Activity and the Growth of the Brain.* Cambridge: Cambridge University
Press, 1994.

Purves, D., and J. W. Lichtman. *Principles of Neural Development.* Sunderland: Sinauer
Associates, Inc., 1985.

Quinn, P. C. "Young Infants' Categorization of Humans Versus Nonhuman Animals:
Roles for Knowledge Access and Perceptual Process." In *Building Object Categories*

in *Developmental Time*, edited by L. Gershkoff-Stowe and D. Rakison, 107–30. Mahwah: Lawrence Erlbaum Associates, in press.

Rage, J. C. "Fossil History." In *Snakes: Ecology and Evolutionary Biology*, edited by R. A. Seigel, J. T. Collins and S. S. Novak, 51–76. New York: Macmillan, 1987.

Resnick, M. *Turtles, Termites, and Traffic Jams: Explorations in Massively Parallel Microworlds*. Cambridge: The MIT Press, 1994.

Richards, R. J. *Darwin and the Emergence of Evolutionary Theories of Mind and Behavior*. Edited by D. L. Hull, *Science and Its Conceptual Foundations*. Chicago: The University of Chicago Press, 1987.

Ridley, M. *Nature Via Nurture: Genes, Experience, and What Makes Us Human*. New York: HarperCollins, 2003.

Robinson, S. R., and G. A. Kleven. "Learning to Move Before Birth." In *Advances in Infancy Research: Prenatal Development of Postnatal Functions*, edited by B. Hopkins and S. Johnson, in press.

Romanes, G. J. *Animal Intelligence*. Washington, D.C.: University Publications of America, Inc., 1883/1977.

Ronca, A. E., and J. R. Alberts. "Effects of Prenatal Spaceflight on Vestibular Responses in Neonatal Rats." *Journal of Applied Physiology* 89 (2000): 2318–24.

Rose, H., and S. Rose, eds. *Alas, Poor Darwin: Arguments against Evolutionary Psychology*. New York: Harmony Books, 2000.

Rosenblatt, J. S. "Daniel Sanford Lehrman: June 1, 1919–August 27, 1972." *Biographical Memoirs of the National Academy of Sciences* 66 (1995): 227–45.

Ruse, M. *Darwin and Design: Does Evolution Have a Purpose?* Cambridge, MA: Harvard University Press, 2003.

Russell, B. *A History of Western Philosophy*. New York: Simon and Schuster, 1945.

Saffran, J. R., R. N. Aslin, and E. L. Newport. "Statistical Learning by 8-Month-Old Infants." *Science* 274 (1996): 1926–28.

Sarkar, S. *Genetics and Reductionism*. Cambridge: Cambridge University Press, 1998.

Schilling, T. H. "Infants' Looking at Possible and Impossible Screen Rotations: The Role of Familiarization." *Infancy* 1 (2000): 389–402.

Schöner, G., and E. Thelen. "Using Dynamic Field Theory to Rethink Infant Habituation." *Psychological Review* (in press).

Scott, J. P., and J. L. Fuller. *Genetics and the Social Behavior of the Dog*. Chicago: Chicago University Press, 1965.

Serpell, J., ed. *The Domestic Dog: Its Evolution, Behaviour, and Interactions with People*. Cambridge: Cambridge University Press, 1995.

Shatz, C. J. "Impulse Activity and the Patterning of Connections During CNS Development." *Neuron* 5 (1990): 745–56.

Shettleworth, S. "Modularity and the Evolution of Cognition." In *The Evolution of Cognition*, edited by C. Heyes and L. Huber, 43–60. Cambridge, MA: MIT Press, 2000.

Shin, T., D. Kraemer, J. Pryor, L. Liu, J. Rugila, L. Howe, S. Buck, K. Murphy, L. Lyons, and M Westhusin. "A Cat Cloned by Nuclear Transplantation." *Nature* 415 (2002): 859.

Simcock, G., and H. Hayne. "Breaking the Barrier? Children Fail to Translate Their Preverbal Memories into Language." *Psychological Science* 13 (2002): 225–31.

Skinner, B. F. *The Behavior of Organisms.* New York: Appleton-Century-Crofts, Inc., 1938.

Slater, A., V. Morison, M. Somers, A. Mattock, E. Brown, and D. Taylor. "Newborn and Older Infants' Perception of Partly Occluded Objects." *Infant Behavior and Development* 13 (1990): 33–49.

Smith, L. B., and E. Thelen. *A Dynamic Systems Approach to Development.* Cambridge: The MIT Press, 1993.

Smith, L. B., E. Thelen, R. Titzer, and D. McLin. "Knowing in the Context of Acting: The Task Dynamics of the a-Not-B Error." *Psychological Review* 106 (1999): 235–60.

Spelke, E. S., K. Breinlinger, J. Macomber, and K. Jacobson. "Origins of Knowledge." *Psychological Review* 99 (1992): 605–32.

Spelke, E. S., and E. L. Newport. "Nativism, Empiricism, and the Development of Knowledge." In *Handbook of Child Psychology. Volume 1: Theoretical Models of Human Development,* edited by W. Damon and R. M. Lerner, 275–340, 1998.

Spencer, J. P., L. B. Smith, and E. Thelen. "Tests of a Dynamic Systems Account of the A-Not-B Error: The Influence of Prior Experience on the Spatial Memory Abilities of Two-Year-Olds." *Child Development* 72 (2001): 1327–46.

Stamps, J. "Behavioural Processes Affecting Development: Tinbergen's Fourth Question Comes of Age." *Animal Behaviour* 66 (2003): 1–13.

Starkey, P., and R. G. Cooper, Jr. "Perception of Numbers by Human Infants." *Science* 210 (1980): 1033–35.

Stent, G. S. "Explicit and Implicit Semantic Content of the Genetic Information." In *Foundational Problems in the Special Sciences,* edited by Butts and Hintikka, 131–49. Dordrecht: D. Reidel Publishing Company, 1977.

———. "Strength and Weakness of the Genetic Approach to the Development of the Nervous System." *Annual Review of Neuroscience* 4 (1981): 163–94.

Taheri, S., J. M. Zeitzer, and E. Mignot. "The Role of Hypocretins (Orexins) in Sleep Regulation and Narcolepsy." *Annual Review of Neuroscience* 25 (2002): 283–313.

ten Cate, C. "Perceptual Mechanisms in Imprinting and Song Learning." In *Causal Mechanisms of Behavioural Development,* edited by J. A. Hogan and J. J. Bolhuis, 116–46. Cambridge: Cambridge University Press, 1994.

Terkel, J. "Cultural Transmission in the Black Rat: Pine Cone Feeding." *Advances in the Study of Behavior* 14 (1995): 119–54.

Thelen, E., and D. M. Fisher. "Newborn Stepping: An Explanation for a "Disappearing" Reflex." *Developmental Psychology* 18 (1982): 760–75.

Thelen, E., G. Schoner, C. Scheier, and L. B. Smith. "The Dynamics of Embodiment: A Field Theory of Infant Perseverative Reaching." *Behavioral & Brain Sciences* 24 (2001): 1–86.

Thompson, R. F. "Behaviorism and Neuroscience." *Behavioral Neuroscience* 101 (1994): 259–65.

Tinbergen, N. "On Aims and Methods of Ethology." *Zeitschrift fur Tierpsychologie* 20 (1963): 410–33.

———. *The Study of Instinct.* Oxford: Oxford University Press, 1951.

Tooby, J., and L. Cosmides. "The Psychological Foundations of Culture." In *The Adapted Mind: Evolutionary Psychology and the Generation of Culture,* edited by J. H. Barkow, L. Cosmides and J. Tooby, 19–136. New York: Oxford University Press, 1992.

Trut, L. N. "Early Canid Domestication: The Farm-Fox Experiment." *American Scientist* 87 (1999): 160–69.

Turati, C. "Why Faces Are Not Special to Newborns: An Alternative Account of the Face Preference." *Current Directions in Psychological Science* 13 (2004): 5–8.

Vandenbergh, J. G. "Prenatal Hormone Exposure and Sexual Variation." *American Scientist* 91 (2003): 218–25.

Venter, J. C., et al. "The Sequence of the Human Genome." *Science* 291 (2001): 1304–51.

vom Saal, F. S., and F. H. Bronson. "Sexual Characteristics of Adult Female Mice Are Correlated with Their Blood Testosterone Levels During Prenatal Development." *Science* 208 (1980): 597–9.

vom Saal, F. S., W. M. Grant, C. W. McMullen, and K. S. Laves. "High Fetal Estrogen Concentrations: Correlation with Increased Adult Sexual Activity and Decreased Aggression in Male Mice." *Science* 220 (1983): 1306–9.

Wahlsten, D. "Behavioral Genetics." In *Encyclopedia of Psychology,* edited by A. E. Kazdin, 378–85. Oxford: Oxford University Press, 2000.

———. "The Intelligence of Heritability." *Canadian Psychology* 35 (1994): 244–58.

———. "The Theory of Biological Intelligence: History and a Critical Appraisal." In *The General Factor of Intelligence: How General Is It?* edited by R. J. Sternberg and E. L. Grigorenko, 245–77. Mahwah, NJ: Lawrence Erlbaum Associates, 2002.

Wakely, A., S. Rivera, and J. Langer. "Can Young Infants Add and Subtract." *Child Development* 71 (2000): 1525–34.

Wallman, J. "A Minimal Visual Restriction Experiment: Preventing Chicks from Seeing Their Feet Affects Later Responses to Mealworms." *Developmental Psychobiology* 12 (1979): 397–97.

Waterland, R. A., and R. L. Jirtle. "Transposable Elements: Targets for Early Nutritional Effects on Epigenetic Gene Regulation." *Molecular and Cellular Biology* 23 (2003): 5293–300.

Watson, J. B. "Psychology as the Behaviorist Views It." *Psychological Review* 20 (1913): 158–77.

Wayne, R. K., and E. A. Ostrander. "Origin, Genetic Diversity, and Genome Structure of the Domestic Dog." *Bioessays* 21 (1999): 247–57.

Weaver, I. C., N. Cervoni, F. A. Champagne, A. C. D'Alessio, S. Sharma, J. R. Seckl, S. Dymov, M. Szyf, and M. J. Meaney. "Epigenetic Programming by Maternal Behavior." *Nature Neuroscience* 7 (2004): 847–54.

Werker, J. F., and A. Vouloumanos. "Speech and Language Processing in Infancy: A Neurocognitive Approach." In *Handbook of Developmental Cognitive Neuroscience*, edited by C. A. Nelson and M. Luciana, 269–80. Cambridge, MA: MIT Press, 2001.

West, M. J., and A. P. King. "Science Lies Its Way to the Truth. Really." In *Handbook of Behavioral Neurobiology*, edited by E. M. Blass, 587–614. New York: Plenum Publishers, 2001.

West, M. J., A. P. King, and T. M. Freeberg. "Social Malleability in Cowbirds: New Measures Reveal New Evidence of Plasticity in the Eastern Subspecies (*Molothrus Ater Ater*)." *Journal of Comparative Psychology* 110 (1996): 15–26.

West-Eberhard, M. J. *Developmental Plasticity and Evolution.* Oxford: Oxford University Press, 2003.

Willis, M. B. "Genetic Aspects of Dog Behaviour with Particular Reference to Working Ability." In *The Domestic Dog: Its Evolution, Behaviour, and Interactions with People,* edited by J. Serpell, 52–64. Cambridge: Cambridge University Press, 1995.

Wilson, E. O. *Sociobiology: The New Synthesis.* Cambridge: Harvard University Press, 1975.

Wyett, W. J., A. Posey, W. Welker, and C. Seamonds. "Natural Levels of Similarities between Identical Twins and between Unrelated People." *Skeptical Inquirer* 9 (1984): 62–66.

Wynn, K. "Addition and Subtraction by Human Infants." *Nature* 358 (1992): 749–50.

———. "Psychological Foundations of Number: Numerical Competence in Human Infants." *Trends in Cognitive Sciences* 2 (1998): 296–303.

# ACKNOWLEDGMENTS

MY INTRODUCTION TO THE CONCEPT of epigenesis began in graduate school at the University of Chicago under the influence of my mentor Howard Moltz, whose respect for T. C. Schneirla and Danny Lehrman was immense. Martha McClintock reinforced the epigenetic perspective and also arranged my introduction to Jeff Alberts in 1983; so off I went to Indiana University where I trained as a developmental psychobiologist. At that time, I was also fortunate to learn from the other members of Indiana's extraordinary group of developmentalists, including the late Esther Thelen.

Many individuals have shaped this book, often in unexpected ways: I thank Liz Freeman for surprising my wife and me with Katy; Mark Rosenthal for the title; and Karen Adolph for "core knowledge." Scott Robinson, Ed Wasserman, and John Freeman, colleagues and friends, are valued resources for stimulating conversation and insight. My students in two seminars on the topic of instinct challenged me to hone my arguments and sharpen my pedagogy.

Many colleagues took the time to answer my questions: I thank

Doug Wahlsten, Jay Rosenblatt, Susan Jones, Gregg Oden, Linda Smith, Stephen Budiansky, Melissa Fleming, and Patrick Bateson. I am also indebted to those who generously agreed to read all or parts of the manuscript and provided helpful comments and encouragement: Jeff Alberts, Scott Robinson, Kelly Mix, Karen Adolph, Greta Sokoloff, John Freeman, Karl Karlsson, Adele Seelke, and Margie Blumberg.

I thank my "rabbi" Steve Bloom for much-needed and well-timed advice and encouragement. Elizabeth Knoll helped me to get this project off the ground. My agent, Ethan Ellenberg, steered me toward John Oakes at Avalon. No one could reasonably ask for wiser counsel.

Writing a book while running a laboratory requires time and support. I thank the National Institutes of Health and my wonderful cadre of students and research assistants for providing both. My family—parents, sisters, and nephews—is a continual reminder to me of how fortunate I am. As is my wife, Jo, whose loving patience never ceases to amaze me.

# INDEX